Springer Theses

Recognizing Outstanding Ph.D. Research

Aims and Scope

The series "Springer Theses" brings together a selection of the very best Ph.D. theses from around the world and across the physical sciences. Nominated and endorsed by two recognized specialists, each published volume has been selected for its scientific excellence and the high impact of its contents for the pertinent field of research. For greater accessibility to non-specialists, the published versions include an extended introduction, as well as a foreword by the student's supervisor explaining the special relevance of the work for the field. As a whole, the series will provide a valuable resource both for newcomers to the research fields described, and for other scientists seeking detailed background information on special questions. Finally, it provides an accredited documentation of the valuable contributions made by today's younger generation of scientists.

Theses are accepted into the series by invited nomination only and must fulfill all of the following criteria

- They must be written in good English.
- The topic should fall within the confines of Chemistry, Physics, Earth Sciences, Engineering and related interdisciplinary fields such as Materials, Nanoscience, Chemical Engineering, Complex Systems and Biophysics.
- The work reported in the thesis must represent a significant scientific advance.
- If the thesis includes previously published material, permission to reproduce this must be gained from the respective copyright holder.
- They must have been examined and passed during the 12 months prior to nomination.
- Each thesis should include a foreword by the supervisor outlining the significance of its content.
- The theses should have a clearly defined structure including an introduction accessible to scientists not expert in that particular field.

More information about this series at http://www.springer.com/series/8790

Sabine Kraushaar

Soil Erosion and Sediment Flux in Northern Jordan

Analysis, Quantification and the Respective Qualitative Impacts on a Reservoir Using a Multiple Response Approach

Doctoral Thesis accepted by
the Martin Luther University Halle-Wittenberg,
Halle, Germany

Springer

Author
Dr. Sabine Kraushaar
Department of Geography and Regional Research
University of Vienna
Vienna
Austria

Supervisor
Prof. Hans-Jörg Vogel
Martin Luther University Halle-Wittenberg
Halle
Germany

The original title of this Thesis is 'Analysis and Quantification of Soil Erosion, Sediment Flux and the Respective Qualitative Impacts on a Reservoir in Northern Jordan Using a Multiple Response Approach'.

ISSN 2190-5053 ISSN 2190-5061 (electronic)
Springer Theses
ISBN 978-3-319-31886-8 ISBN 978-3-319-31888-2 (eBook)
DOI 10.1007/978-3-319-31888-2

Library of Congress Control Number: 2016935606

Printed on acid-free paper

This Springer imprint is published by Springer Nature
The registered company is Springer International Publishing AG Switzerland

Parts of this research have been published in the following journals:

Kraushaar, S., Schumann, T., Ollesch, G., Schubert, M., Vogel, H.-J., Siebert, C., 2015. Sediment fingerprinting in northern Jordan: element-specific correction factors in a carbonatic setting. Journal of Soils and Sediments 15:2155–2173. DOI 10.1007/s11368-015-1179-2.

Kraushaar, S., Ollesch, G., Siebert, C., Vogel, H.-J., Fuchs, M., 2015. Long-term sediment export estimates from Northern Jordan using Roman cisterns as sediment traps. Geoarchaeology 30:369–378. DOI 10.1002/gea.21517.

Kraushaar, S., Herrmann, N., Ollesch, G., Siebert, C., Vogel, H.-J., 2014. Mound measurements—quantifying medium-term soil erosion under olive trees in Northern Jordan. Geomorphology 213:1–12. DOI 10.1016/j.geomorph.2013.12.021.

To those who inspired, payed, critized, supported and will read it. Thank you!

Supervisor's Foreword

The present study focuses on erosion and sediment transport within a catchment Wadi Al-Arab in Northern Jordan—a country ranked as the fourth water-scarce in the world. Besides the loss of soil, erosion leads to an increased sedimentation and pollution of the Wadi Al-Arab water reservoir. The aim of an integrated water management is therefore to improve water infiltration at the soil surface to increase recharge of groundwater resources, and, in turn, to decrease run-off and the transport of sediment loads to surface waters. A fundamental prerequisite to reach these goals is to better understand the mechanisms and local conditions responsible for erosion. This work provides substantial new insight based on a thorough geomorphological study and will be in the position to enhance water availability and quality in a water- and data-scarce country.

The work comprises the identification of the relevant sediment sources throughout the Wadi Al-Arab, the quantification of the erosion processes and an estimation of sediment transport and connectivity to the final sink—the Wadi Al-Arab reservoir. Geochemical analyses of the soil materials in the source regions together with those in the sediments of the reservoir were used in a multiple sediment fingerprinting to calculate the contribution of the various sources. This also provided a substantial understanding of many natural abundant heavy metals and sediment-bound anthropogenic pollutants, such as hexazinone, trifluralin and tNP fertilizer. The results were used to calibrate and validate the SedNet model to obtain a spatial understanding of the processes and to identify hot spot areas. In combination with detailed geomorphological analysis a multiple response approach of different complementary methods was established internalizing the potential of one method validating or falsifying results from another. This led to a consistent interpretation of the ongoing processes.

Overall, this work met the exceptional challenge to identify sediment sources and quantify hydro-morphological processes in a relatively big catchment area of heterogeneous geology, topography and land use. All this was done in the limited time frame of a research project and without support of any pre-existing data. The skillful collection of erosion values in different time- and spatial scales, geochemical properties, and the analysis of sediment connectivity and spatial modelling resulted in an extensive field data set that allowed a thorough understanding of the relevant sediment transport processes. This provides valuable information for policy-makers and resource authorities to prioritize areas suitable for mitigation strategies with respect to erosion and to select possible locations for in-channel constructions to support ground water recharge. I hope this work will stimulate more creative research within regions that are not only scarce in water but also scarce in available data.

Halle, Germany
February 2016

Prof. Hans-Jörg Vogel

Acknowledgments

This dissertation is the result of many thinking heads, numerous helping hands in Germany and in Jordan, caring family members and encouraging friends. These people I want to dedicate the next few lines of sincere appreciation, not ranking them by order and hoping that I haven't forgotten anyone.

First of all I want to express my gratitude to my three supervisors during the years:

- Dr. Christian Siebert who encouraged me to write the project application and supported this work with water chemical expertise and me to keep the focus.
- PD Dr. habil. Gregor Ollesch who always found time for my enquiries and whose comments and ideas gave this thesis the course of action.
- Prof. Dr. habil. Hans-Jörg Vogel who enriched the studies with his comments regarding the small-scale analysis steps and gave me the feeling that I am walking in the right direction.

In Jordan I found great local expertise, support and friendship of Prof. Dr. Elias Salameh and Dr. Marwan Al Raggad. Thank you so much for all your time, cordiality and backing when instruments got stuck in the customs, police wanted proof of the nature of the work, guidance trough Jordan, and overall the inclusion into family and social life by the Al-Raggad and Al-Abdallat family. Without this support I would have given up after one year!

When residing in the research area Wadi Al-Arab Imran and Ferial Melkauwi always gave me a home, moral support, advice with local habits and insights into the daily life of Jordanians. Thank you for this wonderful opportunity to get to know and understand people better.

Additionally I want to express gratitude to Prof. Dr. Dr. Dr. habil. Dieter Viehweger, Dr. Jutta Häser, Dr. Frauke Kenkel and Katja Soennecken from the German Protestant Institute of Archaeology and the Tell Zira'a team. They taught me archaeological working procedures, the ancient history of the region, emphasized integrated working, allowed me to use their infrastructure and more importantly, let me to be a part in their team. Thank you for this support!

The nature of the thesis demanded extensive field work and data analysis, which was strongly supported by the master students, interns and colleagues Cornelia Wilske, Stephan Schulz, Thomas Schumann, Martin Weber and Mascha Braun. Their willingness to accompany and complement the dissertation with their works gave great impulses to the research. Thank you!

Additionally I want to thank Michael von Hoff and Prof. Pöllmann (MLU), from the University of Halle/Wittenberg for grain size and an XRD analysis. Special words of gratitude are devoted to the all technicians at the geochemical laboratories of the TU Dresden and to Mrs. Mothes and the staff of the UFZ Analytic Department in Leipzig for sample analysis, expertise and advice during interpretation.

Furthermore, I am greatly indebted to Manfred Fischer from the luminescence laboratory at the University of Bayreuth for OSL sample preparation and even more to Prof. Markus Fuchs who supported this work with discussions and the OSL dating of the cistern sediments.

I want to thank Oliver Spott and Ralf Gründling for countless hours of fun discussions while commuting back and forth to Halle and this way forgetting about the distance- and work-related problems.

Furthermore, I am very grateful for inclusion, friendship, interest and teachings of Dr. David Morche and Henning Baewert from the University of Halle/Wittenberg and the working group of Young Geomorphologists. Their way of doing team work in science served as a wonderful role model that encourages one to proceed in science when times are rough. Thank you for this collegiality.

Great thanks go to my long university friends André Kirchner and Nico Herrmann whom I owe not only for a hand in the field but also for fruitful discussions before conference presentations and wonderful years of friendship.

To my girls Vicky and Verena, Anne, Marie and Jule, thank you for any kind of positive distraction, lending me your ears and sharing good times.

With all my heart I thank my beloved parents Karl-Heinz and Hannelore, Kitty Emmelmann, my brother Matthias Kraushaar, my husband Kai, Gesine and the Höhnert's and to the extended family. Thank you for giving me the stability, enduring tough times with me, grounding and supporting me at any time in my life!

Last but not least thanks go to the tax payer who gave the money for this work and to the reviewers who spent their time reading it.

Contents

Abbreviations

AD	Anno Domini
a.s.l.	Above Sea Level
ASL/AHP	Amman silicified Limestone/Al-Hisa Phosphorites
Aster	Advanced Spaceborne Thermal Emission and Reflection Radiometer
BGR	Bundesanstalt für Geowissenschaften und Rohstoffe
BMBF	German Federal Ministry of Education and Research
BP	Before Present (1950)
BPA	Bisphenol A
BSh-Csa	Steppe-Mediterranean Climate
CF	Correction Factor
CIA	Central Intelligence Agency
(G)DEM	(Global) Digital Elevation Model
dGPS	Differential Global Positioning System
EC	Electrical Conductivity
EDXRF	Energy Dispersive X-ray Fluorescence
EP	Erosion Pin
FAO	Food and Agriculture Organization
GC MSD	Gas Chromatography coupled with a Mass Spectrometry Detector
GIS	Geographic Information System
GLASOD	Global Assessment of Human-Induced Soil Degradation
GTZ today GIZ	German Society for International Cooperation
HDPE bottles	High-Density Polyethylene bottles
HIGRADE	Helmlholtz Interdisciplinary Graduate School for Environmental Research
HP-5MS	Low bleed column for gas chromatography

HSDR	Hill slope Sediment Delivery Ratio
IAEA	International Atomic Energy Agency
ICP AES	Inductively Coupled Plasma Atomic Emission Spectroscopy
ICP-MS	Inductively Coupled Plasma Mass Spectrometry
IOC	International Oil Council
IP	Influencing Parameters
IPSWAT	International Postgraduate Studies in Water Technologies
ISRIC	International Soil Reference and Information Centre
IUSS Working Group WRB	International Union of Soil Sciences working group for World Reference Base for Soil Resources
IWRM	Integrated Water Resource Management
JDA	Jordan Dam Authority
JICA	Japan International Cooperation Agency
Landsat ETM+ GLS 2010 scenes	Landsat Enhanced Thematic Mapper (Landsat 7) Global Land Survey 2010 scenes
LAWA	Working Group on water issues of the Federal States and the Federal Government
MAD	Median Absolute Deviation
MAR	Managed Aquifer Recharge
MCM	Muwaqqar Chalk Marl
METI	Ministry of Economy, Trade, and Industry (METI) of Japan
MoA	Jordan Ministry of Agriculture
MWI	Jordan Ministry of Water and Irrigation
NASA	National Aeronautics and Space Administration
NDVI	Normalized Difference Vegetation Index
NICE-office	Implementation research office in Jordan
OSL dating	Optically Stimulated Luminescence dating
P	Precipitation
PAHs	Polycyclic Aromatic Hydrocarbons
PET	Potential Evapotranspiration
PPCPs	Pharmaceuticals and Personal care products
RUSLE	Revised Universal Soil Loss Equation
SMART-Project	Sustainable Management of Available Water Resources with Innovative Technologies
SMRAM	Stepwise Multiple Regression Analysis Model
SSA	Specific Surface Area
TAC	Technical Advisory Committee
TIC	Total Inorganic Carbonate

TIN	Triangular Irregular Networks
TOC	Total Organic Carbonate
TrinkwV	German Drinking Water Ordinance
T-value	Soil-loss tolerance value
UN	United Nations
UNEP	United Nation Environment Programme
UNHCR	The UN's refugee agency
URC	Umm Rijam Chert
USEPA	Unites States Environmental Protection Agency
WC-coated	Tungsten Carbide coated
WD	Wadi Deposits
WDXRF	Sequential Wavelength Dispersive X-ray Fluorescence
WHO	World Health Organization

Chapter 1
Introduction

1.1 Background and Framework of the Thesis: The Smart—Project

Jordan is one of the water scarcest countries in the world with the Lower Jordan River Basin constituting an overexploited closed river basin (MWI and GTZ 2004; Venot et al. 2006; Wolf et al. 2007; Hötzl 2004; Hötzl et al. 2009). The physical water scarcity is prompted by an increasing water demand, due to the constantly growing population (3.86 % estimated for 2014, CIA 2014; UNEP 2005). Over the past 20 years the population has doubled, mainly due to refugees from Iraq, Palestine, and Syria which currently make up 34 % of Jordan's population (CIA 2014). The lack of water-related infrastructure and water management has led to an additional qualitative water problem due to leaking septic tanks, lack of waste water treatment facilities, effluents from industry, and diffusive agricultural-related inputs (Abu-Sharar 2006; Ghrefrat and Yusuf 2006), as well as contaminated leachate from landfills (Abu-Rukah and Kofahi 2001) or geologically driven through the dissolution of evaporites (Salameh 2001).

In regard to the worsening situation on site and the millennium development goals (UN resolution 58/217, 2000), Jordan aspires to achieve an Integrated Water Resources Management (IWRM). IWRM describes a process which promotes "the coordinated development and management of water, land, and related resources, in order to maximize the resultant economic and social welfare in an equitable manner without compromising the sustainability of vital ecosystems." (Global Water Partnership 2000).

This implies that suitable strategies, concepts, measures, and technologies are developed especially for a Mediterranean to arid region that result in the optimization of sustainable usage of water resources, despite conceivable climate and land use changes (Geyer and Möller 2011). A consortium of scientists, companies, and authorities of all neighbouring countries to the Jordan River—namely Palestine, Israel, and Jordan plus Germany—set up a research project to deal with this question.

S. Kraushaar, *Soil Erosion and Sediment Flux in Northern Jordan*,
Springer Theses, DOI 10.1007/978-3-319-31888-2_1

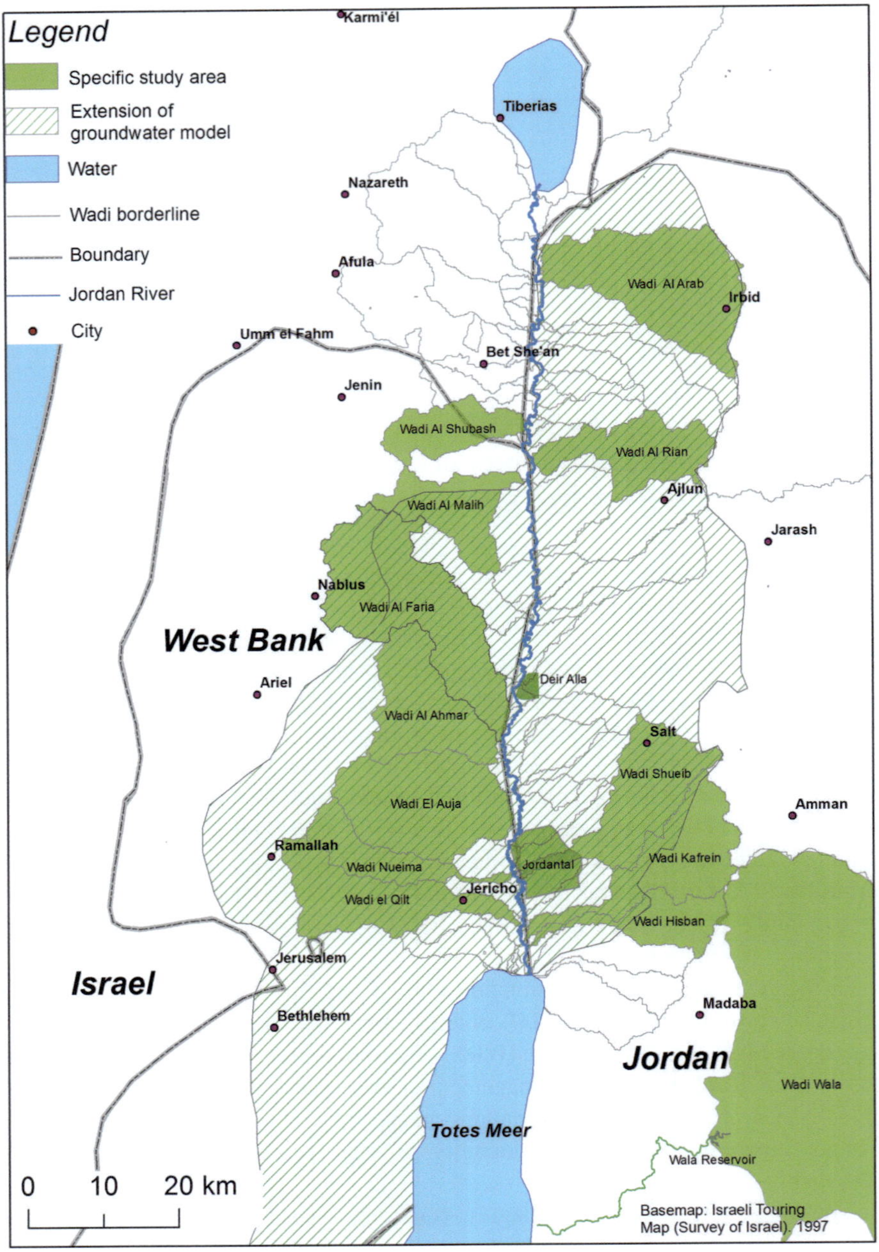

Fig. 1.1 SMART project area—the lower Jordan valley (www.iwrm-smart2.org)

This SMART II Project (= Sustainable Management of Available Water Resources with Innovative Technologies) aims to develop a transferable approach for IWRM in regions of water shortage. The SMART project has been sponsored since 2006 by the German Federal Ministry of Education and Research (BMBF). Activities focus on both sides of the Lower Jordan Rift Valley (Fig. 1.1).

In regard to the water scarcity in Jordan and the three main principles of IWRM: Social equity, economic efficiency, and ecological sustainability, the following questions play a central role in the SMART project:

- How can water availability be increased and water quality improved without endangering vital ecosystems and social and economic welfare?
- Which innovative technologies, decision support systems, and management strategies can be applied in a reasonable and effective way for a sustainable use of water resources?

In the first SMART Phase (2006–2010), the work focused on the investigation and assessment of all utilizable water resources in the region. Results showed that possible methods to encounter water scarcity include, on the one hand, unused water resources as treated waste water, and, on the other hand, the artificial replenishment of aquifers by means of rainwater harvesting (Wolf et al. 2007). In phase II (2010–2013), the identification and characterization of areas with water storage and recovery potential through runoff infiltration are a main focus. This approach intends to reduce overland flow, slow down the runoff to maximize infiltration, and thus artificially recharge the aquifers.

1.2 Demand, Preliminary Works and Research Questions of the Thesis

The soil layer and its physical integrity play a key role in slowing down runoff and safeguarding infiltration, thereby buffering and filtering pollutants (Blum 1997; Pierzynski 1997). The loss of the soil layer on slopes results in the loss of irreplaceable fertile top soil and in decreased infiltration potential, increased overland flow, and accelerated runoff (Symander 2004; Wickel 2007). This again prohibits further infiltration and increases erosion downslope, and hence, augments further soil loss through the cause-effect relationship (Zielhofer and Faust 2002). In conclusion, soil is an important influential parameter to hydrology (Lal et al. 1997; Pierzynski 1997).

In addition to local soil degradation, high rates of sediment loads lead to morphological impacts in riverbeds, sedimentation of reservoirs as well as eutrophication of surface waters (Palmieri et al. 2001; Boardman and Poesen 2006; White 2010). This is of special relevance to Jordan where around 65 % of the freshwater used is derived from surface waters and reservoirs (Nortcliff 2008) which suffer from sedimentation and contamination due to sediment-bound pollutants (Ghrefat and Yusuf 2006; El-Radaideh 2010; Al-Ansari and Shatnawi 2011; Al-Ansari et al. 2012; Ijam and Tarawneh 2012; Renganayaki et al. 2013; Kraushaar et al. 2014).

Any in-channel modifications to enhance infiltration as check dams or recharge basins would have to expect a reduced life span due to high sediment loads in the torrential streams. High sediment loads are described for Wadis from Jordan (Toll and Sauter 2008), and are described as a general problem of the highest sediment yields in catchments with a Mediterranean to semi-arid climate (Vanmaercke et al. 2011).

Hence, when aiming towards an Integrated Water Resource Management, the coordinated management of conserving measurements preventing soil erosion is of great importance in Jordan, with its 35 water reservoirs (MWI data 2011). In this regard, it is important to better understand and quantify erosion and sediment transport. Therefore the exemplary research area Wadi Al-Arab was chosen in the north of Jordan (Fig. 1.1; Chap. 2).

The Wadi Al-Arab catchment lies climatically in the transition zone between the Mediterranean and semi-arid zone (Sect. 2.1), which is an area prone to land degradation in general as reported by many scholars (Langbein and Schumm 1958; Inbar 1992; Kosmas et al. 1997; Conacher and Sala 1998; García-Ruiz et al. 2013). Land degradation according to the United Nation Environment Programme (UNEP 1992) describes the temporary or permanent chemical as well as physical lowering of the productive capacity of land, which, in a semi-arid environment, would be referred to as desertification (FAO 1995). In the following, this work will concentrate on the physical soil degradation denoted to as soil erosion or water erosion throughout the text, as the sediment flux and the qualitative impact is of interest to the research demand.

Reasons for the predisposition of the Mediterranean region in respect to physical erosion of soils are the long history of human activity, decades of intensified agricultural use as well as land abandonment, intense rainstorm events as well as droughts, steep slopes, and the occurrence of forest fires and clearance of vegetated strips (Kosmas et al. 1997; Cerdà 2002; Klewinghaus et al. 2006; García-Ruiz 2010; García-Ruiz et al. 2013). In this regard, especially marly areas have been extensively agriculturally used throughout history, resulting in strong erosive features (Faust et al. 2009; Faust and Schmidt 2009; Casalí et al. 1999; Cerdà 2002). In Jordan, a first mapping and assessment of soil degradation was implemented within the global efforts of the UNEP and the International Soil Reference and Information Centre (ISRIC 1990). In the framework of the "Global Assessment of Human-Induced Soil Degradation" (Project GLASOD), a high degree of water and wind erosion (ISRIC 1990; FAO 2009; Richter 1998) was identified on a very broad analytical scale, covering the whole earth. Khresat et al. (1998) summarize water and wind erosion as the most prominent processes leading to soil erosion in northern Jordan. According to the authors, the processes are mainly controlled by overgrazing, conversion of rangelands to croplands in marginal areas, and improper farming practices (Khresat et al. 1998). Abu-Sharar (2006) describes the latter in more detail as uncontrolled herd movements, firewood collection and hence, deforestation, ploughing down slopes, and the use of heavy farm machinery.

He adds that periodic draughts, the high population growth rate, and weak institutional arrangements that promote soil erosion constantly lower the land productivity.

Schmidt et al. (2006) and Lucke (2007) analysed whether historical enhanced soil erosion led to an abandonment of northern Jordan in former times (ca. 1200 BP). Lucke concluded that certainly the long-term land use impacted the landscape development but did not cause major historic erosion, as most colluvium date back to the transition phase of the Pleistocene to the Holocene (Lucke 2007, 2008). Furthermore, through the analysis of soil profiles, he proved that the aeolian input has been minimal since the abandonment of the area around 1200 years ago, which is supported by the findings of Jahn (1995) who detected only two percent aeolian input in recent soils in the region (Jahn 1995; Lucke 2007). Hence, it can be assumed, that the recent addition of allochthonous material is marginal. Furthermore, the very clayey (ø = 29.4 %, n = 262 samples, analysis by the ML University of Halle/Wittenberg, M. v. Hoff) substrate in the mountainous area as the Wadi Al-Arab catchment is not suitable for aeolian transport as it exists in polyedric or subpolyedric aggregates of >2 mm in size. Own observations of the catchment showed mainly water erosive rather than aeolian features (Chap. 3). These findings led to the decision to focus exclusively on the dominant processes of water erosion in this thesis, although the author cannot exclude minor wind erosion processes absolutely.

Most of the few erosion studies from Jordan use solely a GIS based (Revised) Universal Soil Loss Equation ((R)USLE) approach to estimate the spatial patterns of erosion (Al-Zitawi 2006; Al-Alawi and Abu Jamous 2009; Farhan and Al-Bakri 2012; Farhan et al. 2013, 2014). Due to the lack of appropriate data in the region, the modeling approach allows to consider the problem only on a very broad spatial scale. The studies usually lack ground truthing and validation of the results with measurements and observations. Modeled results are, when discussed, only on the basis of literature findings from around the world, which when considering the variety of measurement methods and scales always find comparative values (García-Ruiz et al. 2013). Qualitative impacts of the sediments on water bodies cannot be assessed using an empirical hillslope erosion model and thus, are just anticipated, resulting in wrong conclusions (Ghrefat and Yusuf 2006, Chap. 4). Furthermore, the issue of sediment connectivity between the slopes and the riverbed is not taken into account. Thus, it can only be speculated if the eroded sediments from the different land use units are those that contaminate and clog the water course. Other sources, such as paleo terraces along the riverbed, bank erosion, gullies, and "new sediments" from road construction, leaving masses of unconsolidated bedrock sediments near roads, are as well possible major contributors.

In the light of such limited erosion studies, García-Ruiz et al. (2013) identify the future challenges of soil erosion studies in the Mediterranean. The authors argue to perceive soil erosion as a complex system, which needs a holistic approach. More precisely, they suggest amongst others, to improve long-term quantitative information on soil erosion and to develop field methods which improve the identification of sediment sources (García-Ruiz et al. 2013).

This thesis aims to meet the demand in Jordan to better understand erosion and sediment flux in the exemplary catchment Wadi Al-Arab, while at the same time implement and contribute to the development of methods, that give long term average erosion rates and geochemically identify sediment sources.

The objectives of the research are to identify and quantify dominant erosion processes and affected areas. Furthermore, the work analyses sediment transport and sedimentation throughout the water course, as well as the qualitative impact on the Wadi Al-Arab reservoir. Results enable to focus erosion mitigation strategies on areas with high impact on sediment yield. Suitable river stretches for in channel modifications, to reduce runoff and enhance infiltration, are identified.

The objectives are operationalized in the following research questions:

1. Is the Wadi Al-Arab susceptible to soil erosion? If yes, where does erosion take place and in what range does it affect different landscape units?
2. What are the consequences of erosion and sedimentation in regard to the sediment yield, reservoir storage, and water quality?
3. What is the spatial extent of erosion and where are the erosion hot spots? How good is the sediment connectivity in the catchment? How sensitive is the system to changes in land use?

1.3 Methodological Approach and the Structure of the Thesis

The region is characterized by data scarcity to the extent where information lack in an appropriate density and quality, and basic information, such as a parameterized soil map, is non-existent. In respect to this absence of data and the future challenges of soil erosion studies as defined by García-Ruiz et al. (2013), a multiple response approach was chosen. It combines different methods of erosion measurements on slopes with a detailed analysis of the sink. Furthermore, a Sediment Fingerprint study of the reservoir sediments allows the geochemical identification of the sources and the calculation of their relative contribution in the sink. Results of the measurements and the Sediment Fingerprint cover different spatial scales and partly integrate the transport route. Thus, although they cannot be compared directly, some areas of intersection exist, where findings of one method can validate or falsify results from another. In the optimal case, outcomes that are independently supported by different methods are believed to approximate the truth the most. And are used in the final step to calibrate and validate the setup of the SedNet model (Wilkinson et al. 2008; Fig. 1.2).

The thesis is presented as monograph with a structure similar to Fig. 1.2. After an introduction to the natural characteristics of the research area (Chap. 2), a short overview is given on the major land use units, in which erosion measurements were implemented (Chap. 3). In Sect. 3.1 the use of Roman cistern sediment archives for

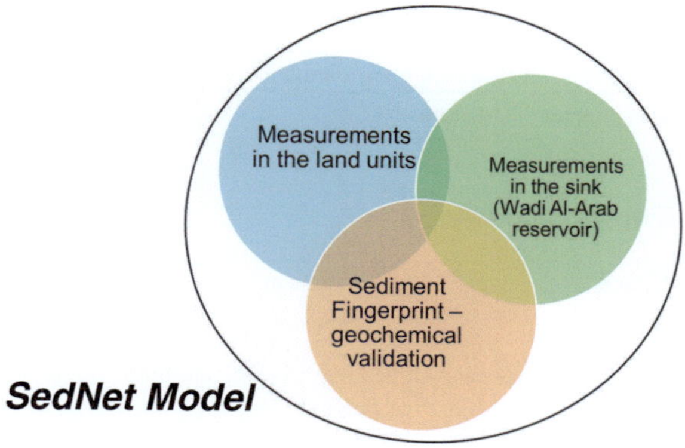

Fig. 1.2 Multiple response approach to encounter the question of soil erosion and sediment transport in the research area

the evaluation of erosion values of levelled positions is explained. Section 3.2 consists of erosion estimation in olive orchards on steep slopes (>10 %). Measurements of micro topographic forms under olive trees, namely olive mounds, and dendrochronological age estimations allowed evaluating the mean yearly soil loss. The following chapters give a short explanation of the use and results of erosion pin fields and recent sediment traps in the Wadi Al-Arab (Sects. 3.3 and 3.4).

The results of the sink and water analysis are presented in Chap. 4. Effects of the sediment load on the reservoir storage and its' water quality are discussed.

The geochemical Sediment Fingerprint is described in Chap. 5. The method helped to identify, differentiate and quantify the relative contribution of 6 sediment sources. The implementation of the method led to developments of correction factors and an adapted selection process for conservative elements.

Chapter 6 merges the findings with the setup of the SedNet model. Areas of high erosion and sediment connectivity rates are discussed in regard to possible management strategies and in in channel modification rendering the discharge.

Each chapter can stand alone in its structure and is devoted to a different part of the research questions. A state-of-the-art discussion is always given in the beginning. Methods and results are elaborated and discussed in each chapter independently. Finally, all the findings are summarized (Chap. 7), the outlook presented (Chap. 8) and innovative aspects of the research mentioned (Chap. 9).

It is believed that the structure of completed storylines, which are merged in the end, allows the reader to follow the diverse methods and research questions better, without getting lost in a "forest of method description and listings of results". A colour-coded assembly of the research questions with their respective chapters is given in Table 1.1. The colours represent their part in the multiple response approach as seen in Fig. 1.2.

Table 1.1 Overview of the thesis structure and chapters

Chapter	Research question/intent	Content
This chapter	Structure	Framework of the thesis, research question and objectives as well as the structure
Chapter 2	Research area	An introduction to the natural characteristics of the research area
Chapter 3	Is the Wadi Al-Arab susceptible to soil erosion? If yes, where does erosion take place and in what range does it affect different landscape units?	3.1: Long-Term Sediment Export Estimates from Northern Jordan using Roman Cisternsas Sediment Traps 3.2: Mound measurements — quantifying medium-term soil erosion under olive trees in Northern Jordan 3.3: Implementation and results of 6 erosion pin fields 3.4: Building of event based sediment traps and the assessment of a few events
Chapter 4	Consequences for the reservoir and water quality?	Results of the Wadi Al-Arab reservoir analysis and water quality estimations
Chapter 5	How good is the sediment connectivity in the catchment? Or in other words: Are the sediments in the reservoir eroded from the slopes in recent years or from intermediate sediment storages?	Sediment fingerprinting in northern Jordan: element-specific correction factors in a carbonatic setting
Chapter 6	What is the spatial extent of erosion and where are the erosion hot spots? How sensitive is the system to changes in land use?	Wrapping up findings in a sediment transport model for North Jordan
Chapter 7	Wrap up of the results	Conclusion and outlook of the results

References

Abu-Rukah Y, Al-Kofahi O (2001) The assessment of the effect of landfill leachate on ground-water quality—a case study. El-Akader landfill site—north Jordan. J Arid Environ 49:615–630

Abu-Sharar TM (2006) The challenges of land and water resources degradation in Jordan: diagnosis and solutions. In: Kepner WG, Rubio JL, Mouat DA, Pedrazzini F (eds) Desertification in the Mediterranean. Springer, Netherlands, pp 201–226

Al-Alawi M, Abu Jamous M (2009) Estimation of soil erosion in Jordan by using GIS. In: De Amicis R et al (eds) GeoSpatial visual analytics: geographical information processing and visual analytics for environmental security. Springer, Berlin, pp 439–450

Al-Ansari N, Shatnawi A (2011) Siltation of three small reservoirs in Jordan. Publications by the Institute of Earth and Environmental Sciences, Al-Bayt University Jordan

Al-Ansari N, Alroubai A, Knutsson S (2012) Bathymetry and Sediment survey for two old water harvesting schemes, Jordan. J Earth Sci Geotech Eng 2(2):13–23

Al-Zitawi FI (2006) Using RUSLE in prediction of soil loss for selected sites in north and north west of Jordan. M.Sc thesis. Jordan University of Science and Technology, Irbid

Blum WEH (1997) Basic concepts: degradation, resilience and rehabilitation. In: Lal R, Blum WH, Valentine C, Stewart BA (eds), Methods for assesment of soil degradation. Boca Raton, New York

Boardman J, Poesen J (eds) (2006) Soil erosion in Europe. Wiley, Chichester

Casalí J, López JJ, Giráldez JV (1999) Ephemeral gully erosion in southern Navarra, Spain. Catena 36:65–84

Cerdà A (2002) The effect of season and parent material on water erosion on highly eroded soils in eastern Spain. J Arid Environ 52:319–337

CIA (2014) The world factbook—Jordan. https://www.cia.gov/library/publications/the-world-factbook/geos/jo.html. Last checked 08 Apr 2014)

Conacher AJ, Sala M (eds) (1998) Land degradation in Mediterranean environments of the world. Nature and extent, causes and solutions. Wiley, Chichester/England

El-Radaideh NM (2010) Using bottom reservoir sediments as a source of agricultural soil: Wadi El-Arab reservoir as a case study, NW Jordan. Abhath Al-Yarmouk: Basic Sci Eng 19(2): 75–91

Farhan IA, Al-Bakri FT (2012) Use of GIS and remote sensing to assess soil erosion in an arid to semiarid basin in Jordan. In: Proceedings of the international conference on sediment transport: modeling in hydrological watersheds and rivers. Turkey, Istanbul, pp 145–152

Farhan Y, Zregat D, Farhan I (2013) Spatial estimation of soil erosion risk using RUSLE approach, RS, and GIS techniques: a case study of Kufranja watershed, Northern Jordan. J Water Resour Prot 5(12):1247–1261

Farhan Y, Zregat D, Nawaiseh S (2014) Assessing the influence of physical factors on spatial soil erosion risk in northern Jordan. J Am Sci 10(7):29–39

Faust D, Schmidt M (2009) Soil erosion processes and sediment fluxes in a mediterranean marl landscape, Campiña de Cádiz, SW Spain. Z für Geomorphol 52:247–265

Faust D, Heckmann T, Moldenhauer KM, Schütt B, Zielhofer C (2009) Klimawandel – Bedrohung für die Erdoberfläche. In: Deutscher Arbeitskreis für Geomorphologie (eds) Denkschrift 2: Die Erdoberfläche – Lebens- und Gestaltungsraum des Menschen. Beiträge der Geomorphologie zur Erforschung des globalen Wandels, pp 3–15

Food and Agriculture Organization of the UN (FAO) (1995) Land degradation in south Asia: its severity, causes and effects upon the people. http://www.fao.org/docrep/v4360e/v4360e03.htm. Last checked 2 Jun 2014

Food and Agriculture Organization of the UN (FAO) (2009) Jordan-severity of human induced soil degradation http://www.fao.org/landandwater/agll/glasod/glasodmaps.jsp?country=JOR&-search=Display+map+! Last checked 04 Aug 2009

García-Ruiz JM (2010) The effects of land uses on soil erosion in Spain: a review. Catena 81:1–11

García-Ruiz JM, Nadal-Romero E, Lana-Renault N, Beguería S (2013) Erosion in Mediterranean landscapes: changes and future challenges. Geomorphology 198:20–36

Geyer S, Müller R (eds) (2011) IWRM- SMART Teilprojekt 2, Endbericht. Helmholtz Zentrum für Umweltforschung – UFZ, Halle/Leipzig

Ghrefat H, Yusuf N (2006) Assessing Mn, Fe, Cu, Zn, and Cd pollution in bottom sediments of Wadi Al-Arab Dam, Jordan. Chemosphere 65:2114–2121

Global Water Partnership, Technical Advisory Committee (TAC) (2000) Integrated water resource management, TAC background papers no. 4. Global Water Partnership, Sweden, pp 8–50

Hötzl H (2004) Natural scarcity of water resources in the semi-arid and arid middle east and its economical implications. In: Zereini F, Jaeschke W (eds) Water in the middle east and in North Africa. Springer, Berlin

Hötzl H, Guttman J, Salameh E, Tamimi AR (2009) State of water strategy and policy. In: Hötzl H, Möller P, Rosenthal E (eds) The water of the Jordan Valley. Springer, Berlin

Ijam AZ, Trawneh ER (2012) Assessment of sediment yield for Wala dam catchment area in Jordan. Eur Water 38:43–58

Inbar M (1992) Rates of fluvial erosion in basins with a mediterranean type climate. Catena 19:393–409

International Soil Reference and Information Centre (ISRIC) (1990) http://www.isric.org/projects/global-assessment-human-induced-soil-degradation-glasod. Last checked 01 April 2016

Jahn R (1995) Ausmaß äolischer Einträge in circumsaharische Böden und ihre Auswirkungen auf Bodenentwicklung und Standortseigenschaften. Hohenheimer Bodenkundliche Hefte 23

Jordan Ministry of Water and Irrigation (MWI) & GTZ (2004) National water master plan of Jordan

Khresat SA, Taimeh AY (1998) Properties and characterization of vertisols developed on limestone in a semi-arid environment. J Arid Environ 40:235–244

Khresat SA, Rawajfih Z, Mohammad M (1998) Land degradation in north-western Jordan: causes and processes. J Arid Environ 39:623–629

Klewinghaus A, Turkelboom F, Skowronek A (2006) A GPS/GIS-integrated approach to the assessment of current soil erosion by water—experiences from Mediterranean NW-Syria. Z für Geomorphol 142:281–305

Kosmas C, Danalatos N, Cammeraat LH, Chabart M, Diamantopoulos J, Farand R, Gutierrez L, Jacob A, Marques H, Martinez-Fernandez J, Mizara A, Moustakas N, Nicolau JM, Oliveros C, Pinna G, Puddu R, Puigdefabregas J, Roxo M, Simao A, Stamou G, Tomasi N, Usai D, Vacca A (1997) The effect of land use on runoff and soil erosion rates under Mediterranean conditions. Catena 29:45–49

Kraushaar S, Herrmann N, Ollesch G, Vogel H-J, Siebert C (2014) Mound measurements—quantifying medium term soil erosion under olive trees in Northern Jordan. Geomorphology 213:1–12

Lal R, Blum WH, Valentine C, Stewart BA (eds) (1997) Methods for assesment of soil degradation. CRC Press, Boca Raton

Langbein WB, Schumm SA (1958) Yield of sediment in relation to mean annual precipitation. Am Geophys Union 39(6):1076–1084

Lucke B (2007) Demise of the decapolis. Past and present desertification in the context of soil development, land use and climate. Dissertation at the Brandenburg Technical University of Cottbus

Lucke B (2008) Wasser oder Boden? Welches war der Schlüssel für die Dekapolis Region (Nordjordanien). In: Ohlig C (ed) Cura Aquarum in Jordanien. - Schriften der Deutschen wasserhistorischen Gesellschaft E.V., Band 12, pp 177–188

Ministry of Water and Irrigation (MWI), Jordan Valley Authority (JVA) (2011) Wild plants in Jordan valley. Water Information and Public Awareness Department, Jordan

Nortcliff S, Carr G, Potter RB, Darmame K (2008) Jordan's water resources: challenges for the future. Geograph Paper 185:1–24

Palmieri A, Shah F, Dinar A (2001) Economics of reservoir sedimentation and sustainable management of dams. J Environ Manage 61(2):149–163

Pierzynski GM, Hettiarachchi GM, Koelliker JK (1997) Methods of assessing the impacts of soil degradation on water quality. In: Lal R, Blum WH, Valentine C, Stewart BA (eds) Methods for assessment of soil degradation. CRC Press, New York

Renganayaki SP, Parimala S, Elango L (2013) A review on managed aquifer recharge by check dams: a case study near Chennai, India. Int J Res Eng Technol 2(4):416–423

Richter G (1998) Bodenerosion als Weltproblem. In: Richter G (ed) Bodenerosion – Analyse und Bilanz eines Umweltproblems. Wissenschaftliche Buchgesellschaft, Darmstadt

Salameh E (2001) Sources of water salinities in the Jordan valley area/Jordan. Acta Hydrochim Hydrobiol 29(6–7):329–362

Schmidt M, Lucke B, Bäumler R, Al-Saad Z, Al-Qudah B, Hutcheon A (2006) The decapolis region (Northern Jordan) as historical example of desertification? Evidence from soil development and distribution. Quatern Int 151:74–86

Symander W (2004) Was passiert, wenn der Regen fällt? Verlag Eugen Ulmer, Stuttgart, Einführung in die Hydrologie

Toll M, Sauter M (2008) Rainfall-runoff relationships in the area of Wadi Kafrein, Jordan. In: Wolf L, Pöschko A, Werz H (eds) SMART-sustainable management of available resources with innovative technologies. Project Status Report, Karlsruhe, pp 110–112

UNEP, Plan Bleu (2005) A sustainable future for the mediterranean: the blue plan's environment and development outlook. Earthscan, London

United Nations Environment Programme (UNEP) (1992) World atlas of desertification, volume 3/4. London

Vanmaercke M, Poesen J, Verstraeten G, de Vente J, Ocakoglu F (2011) Sediment yield in Europe: spatial patterns and scale dependency. Geomorphology 130:142–161

Venot J-P, Molle F, Courcier R (2006) Dealing with closed basins: the case of the Lower Jordan River Basin. World Water Week 2006, Stockholm

White WR (2010) World water: resources, usage and the role of man-made reservoirs. Foundation of Water Research, Marlow

Wickel AJ, van de Giesen NC, SÁ TDDA (2007) Stormflow generation in two headwater catchments in eastern Amazonia, Brazil. Hydrol Process 22(17):3285–3293

Wilkinson S, Henderson A, Chen Y, Sherman B (2008) SedNet user guide. Client report, CSIRO land and water, Canberra

Wolf L, Werz H, Hoetzl H, Ghane M (2007) Exploring the potential of managed aquifer recharge to mitigate water scarcity in the lower Jordan River basin within an IWRM approach. In: Fox P (ed) Management of aquifer recharge for sustainability. Proceedings of the 6th international symposium on managed artificial recharge of ground water, ISMAR6, Phoenix, Arizona/USA, October 28–November 2

Zielhofer C, Faust D (2002) Mediterranes Klima und Mensch-Umwelt-Beziehungen. Praxis Geographie 3:10–14

Chapter 2
The Wadi Al-Arab

The research area Wadi Al-Arab is located in northern Jordan on the border of Syria and Israel, extending from E 747000 to E 770000 and N 3596000 to N 3618000 (UTM, WGS84, zone 36N) and covering an area of approximately 263.5 km^2. In the north, the catchment is confined by the Yarmouk river, to the south the Ajloun mountains, to the east by Wadi Shellal, and the Jordan River to the west (Fig. 2.1).

The catchment was chosen for the following reasons:

- It is similar to many in the Lower Jordan Valley and thus exemplary for the region. It descends from the Jordan Valley Escarpment down to the Jordan River. The elevation ranges from more than 500 m in the east to around −164 m a.s.l. at the Wadi Al-Arab dam.
- It has an appropriate size of 263 km^2, small enough to get to know most of the area in three years of research, and of a suitable size to evaluate the human impact on erosion while covering major land use units (personal communication Prof. Dr. Faust, EGU, 2013).
- The Wadi Al-Arab is originally a tributary to the Jordan River that is dammed before reaching the valley. The reservoir is one of the oldest in Jordan and has functioned as a final sink for the transported sediments since it came into operation in 1986. This is a great advantage for total sediment volume calculations and the sediments in the reservoir have served as an sediment archive since completion.
- To the east the catchment shows an agricultural basin. From here to the north, northwest, and southwest the relief energy increases until the Wadi Al-Arab reservoir. Its elevation ranges from more than 500 m a.s.l. in the east to around −164 m a.s.l. at the Wadi Al-Arab dam. Different land uses such as wheat and vegetable fields, tree orchards (olives, almonds), and grazing areas reflect the commonly distributed agricultural possibilities of Jordan.
- The City of Irbid is situated in the east as the third biggest agglomeration in Jordan with around 300,000 inhabitants. This has the advantages that the catchment is not only a pristine natural catchment scarcely inhabited but reflects

S. Kraushaar, *Soil Erosion and Sediment Flux in Northern Jordan*,
Springer Theses, DOI 10.1007/978-3-319-31888-2_2

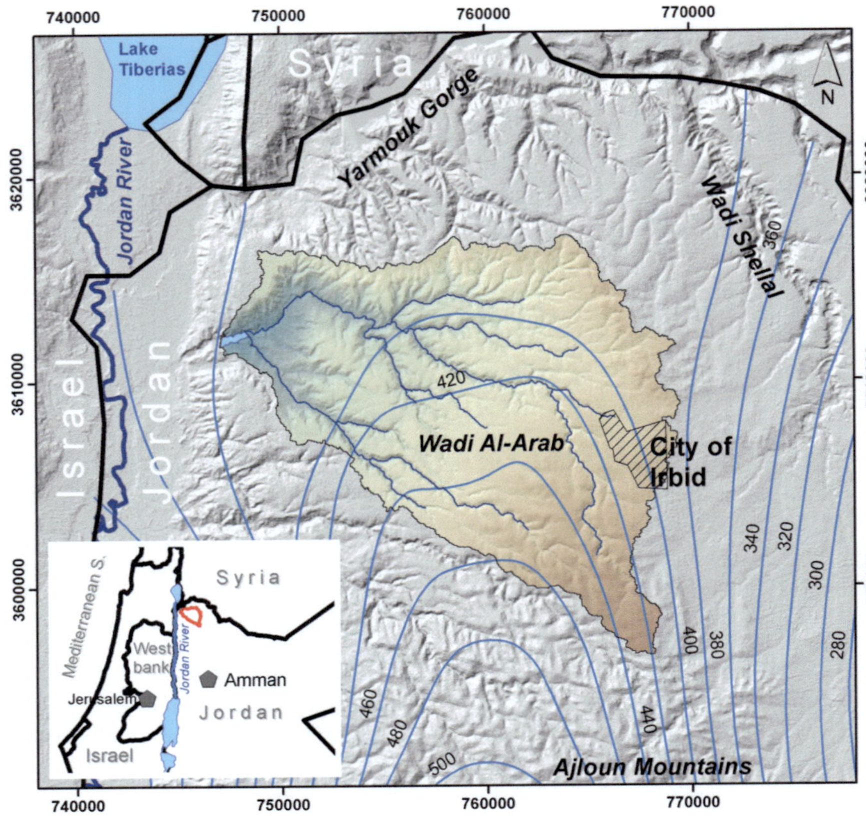

Fig. 2.1 Location of Wadi Al-Arab (MWI data 2010)

the population pressure, agricultural use, street density etc. that is characteristic of many areas along the Jordan River. Last but not least, field installations, special craftsmen, and workshops as well as extraordinary procurements demanded the possibility of a bigger city nearby.

The following section give a more detailed insight into its climate, geology, geomorphology, soil as well as vegetation, and land use.

2.1 Climate in Wadi Al-Arab

The climate in Wadi Al-Arab is classified as BSh-Csa climate according to the updated world Köppen-Geiger Classification (Peel et al. 2007). BSh describes a steppe climate where precipitation is below the potential evapotranspiration, average temperature above 18 °C, and the coldest month above 0 °C. Whilst Csa

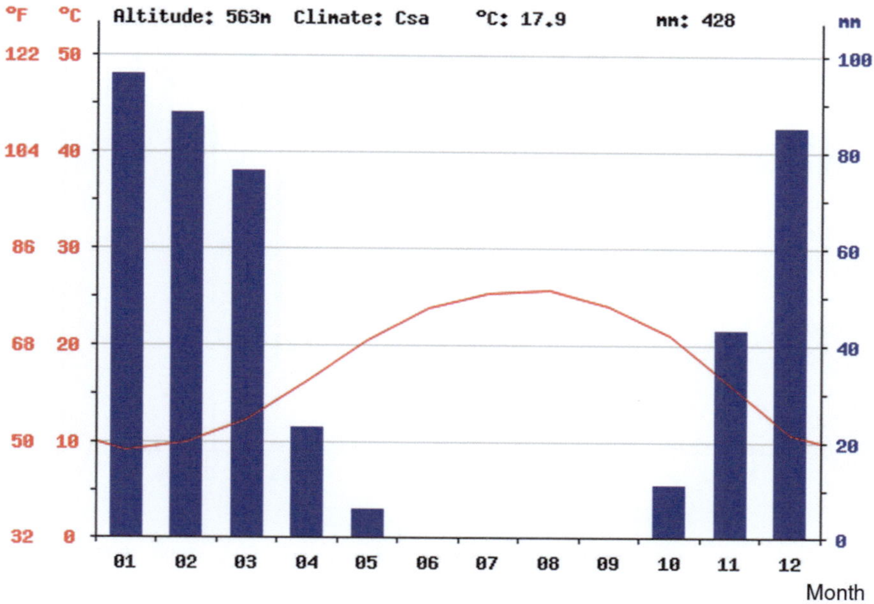

Fig. 2.2 Climate graph Irbid (http://en.climate-data.org/location/3388/)

characterizes a Mediterranean climate with average temperatures above 10 °C, dry summers and the warmest month are above 22 °C.

Rain only occurs between September and May, resulting in an average yearly rainfall of about 495 mm in the south, to around 390 mm in the west (MWI data 2007; Fig. 2.2). Potential evapotranspiration is low in the SE with 1312.75 mm $year^{-1}$ and rises to 1425.44 mm $year^{-1}$ in the NW of the catchment. The resulting aridity index (P/PET; UNEP 1992) describes the catchment with values between 0.27 and 0.35 as semi-arid.

Figure 2.3 indicates that between 1990 and 2007, the first decade (1990–1999) had on average higher rainfall compared to the years 2000–2007. A high inter-annual variability of the rainfall distributions is visible over the years. Stronger rain events were reported and are documented by the rain stations for the years 1992, 1997, and 2003. Besides the usual yearly flash floods, these events showed an enormous erosive potential as reported by farmers in the region.

The analysis of daily rainfall data and discharge volume allowed the calculation of a runoff coefficient. Most pronounced rainfall events show a runoff coefficient below 10 %, similar to those found by Gómez et al. (2009) in southern Spain (12 %).

Pronounced events or when several days of rain occur in a row, values of up to 30 % are possible (1992, 1996, 1997, 1998, 2003; Fig. 2.3). As most rain events lead to minimal discharge the average is 1.6 %.

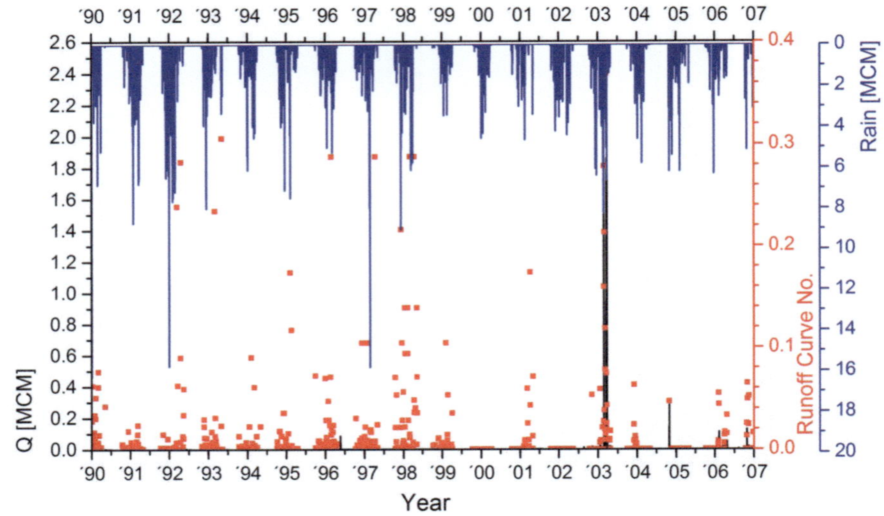

Fig. 2.3 Discharge (Q), rainfall, and calculated runoff curve number for 1990–2007 (JDA data 2013; MWI data 2013)

Eltaif et al. (2010) estimate the kinetic energy of rainfall events in northern Jordan using the RUSLE (Renard et al. 1997) between 200 and 400 MJ mm $ha^{-1}h^{-1}year^{-1}$ and conclude that the values are low compared to humid areas. However, other factors such as the sparse or missing vegetation cover and soil layer reduce the retention potential in the catchment and lead, together with pronounced rainfall events of high intensities, to torrential runoff events known as flash floods. These short-term peak discharges are characteristic of the region and result in a high erosive impact on slopes and as well as in the riverbed, again augmenting erosion following the cause-effect principle (Zielhofer and Faust 2002; Eltaif et al. 2010).

2.2 Geology of Wadi Al-Arab

This chapter will first give a brief introduction to the geological history and the genetic environment of the dominant geological units in the catchment, and then it will focus on north Jordan. For more detailed information on the geological genesis of the region please refer to Bender (1968), Horowitz (2001), and others.

2.2.1 Brief Geological History of the Middle East

In the Precambrian, during the Pan African Orogenesis, the region today known as the Middle East was formed as a stable continental margin (Siebert 2005). In the following pre-Neogene the region was situated in the periphery of the continental shelf and continuously "shaped and controlled by the struggle between the Arab Nubian Massif to the south and the Tethys Ocean to the north" (Flexer 2001). The marine trans- and regression of the Tethys-Ocean and the Arab Nubian Massif as "stable nucleus" dictated the geological evolution of the surrounding areas with the accumulation of shallow marine sediments and the delivery of terrestrial sediments (Flexer 2001; Moh'd 2000). Up until the Cretaceous, a gradual uplifting of the region occurred together with strong formative tectonic phases.

The arching of the Arab Nubian Massif from the Eocene onwards steadily led to the spreading of the region today known as the Red Sea and the Gulf of Aden. Contemporaneously the plate boundaries between the Sinai and the Arabian plate and the Dead Sea—Jordan Rift System developed (Siebert 2005).

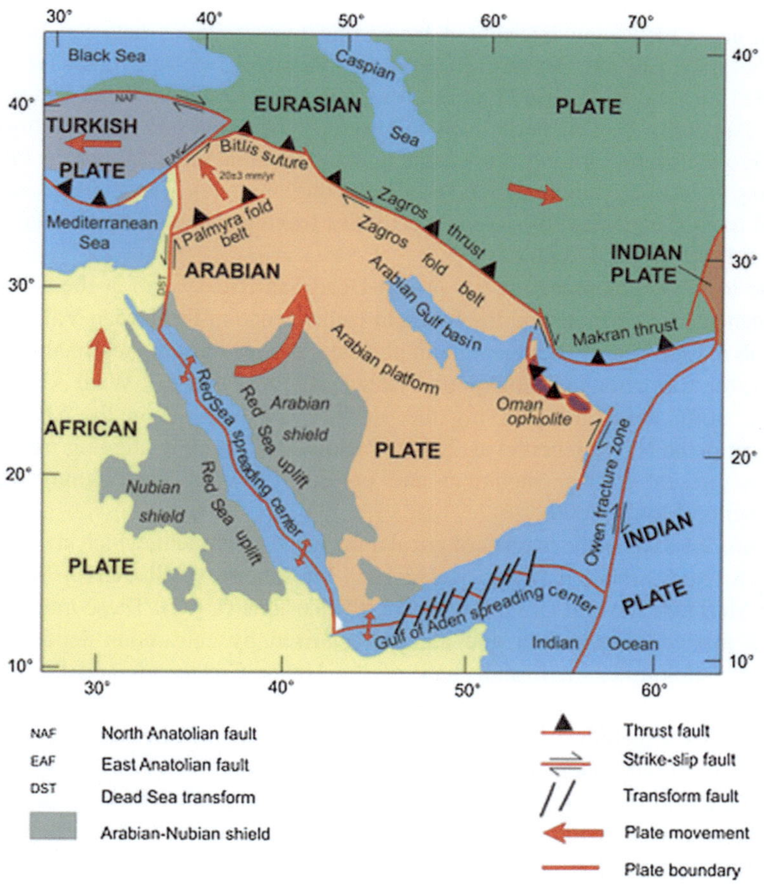

Fig. 2.4 Plate tectonics in the Middle East (Horowitz 2001; http://iv-g.livejournal.com/295425.html)

The Middle East has been situated at this conjunction zone of the African, Arabian, and Anatolian plates, respectively, leading to ongoing tectonic processes in the region. Consequently, the region has always been affected by varying stress fields including spreading, divergence, and shearing along the Red Sea, the Cyprian Arc and the Taurus mountain range and the Dead Sea-Jordan Rift Valley, respectively (Fig. 2.4; Flexer 2001).

2.2.2 *Environmental Conditions for the Geology in Wadi Al-Arab*

Putting the focus on NW Jordan, Wadi Al-Arab as part of the "Ajloun plateau" developed during the Upper Cretaceous, when Jordan was covered by the Tethys Ocean. The sedimentation environment was dominated by shallow tidal-lagoonal conditions, which were interrupted by several transgressions between Santonian and the Late Eocene, leaving shallow to moderately pelagic chalks (Amman Silicified Limestone/Al Hisa Phosphorites, Muwaqqar Chalk Marl and Umm Rijam Chert; Appendix 2). This period was followed by the regression of the Tethys until the Oligo-/Miocene forming a limestone unit and the final regression in the Pliocene allowing lacustrine sedimentation (Waqqas Conglomerate). The more recent geological history is marked by volcanic activities leaving sheets of basaltic lava from about 5.1 Ma (Moh'd 2000).

Due to the development of the Jordan-Dead Sea Rift system at the end of the Oligocene/early Miocene and the following subsidence of the Jordan Valley, Wadi Al-Arab incised deeply into the Ajloun plateau. As a result of the regional structural history, the formations of the Ajloun Plateau including those of Wadi Al Arab are dipping north westwards, leaving the oldest rocks exposed in the SE and the youngest in the NW (Siebert et al. 2014). Various faults mainly aligning from NE to SW are present in the catchment and provide proof of the ongoing tectonic movements (Moh'd 2000).

Figure 2.5 depicts the main geological units in the catchment, which start with the oldest Amman Silicified Limestone/Al Hisa Phosphorite (ASL/AHP), Muwaqqar Chalk Marl (MCM) and Umm Rijam Chert Limestone (URC). These units derived from a marine environment and are characterised by calcareous depositions as limestone, dolomite, marl, and chalk of the Upper Cretaceous and the Tertiary interrupted by chert facies and siliclastics (Moh'd 2000). Especially the Al Hisa Phosphorites, which make up the 8–10 m of phosphoric rock between the ASL and MCM, have a high content of Uranium, ranking Jordan number 11 on the international scale of uranium deposits with locally up to 800 ppm Uranium (Bossone et al. 2013; Moh'd and Powell 2010). The genesis of that stratum is under debate, ranging from authigenic precipitation from seawater to synsedimentary phosphatization as a diagenetic or microbial process (Abed et al. 1989; Abed and Fakhouri 1990; Al-Sharhan and Nairn 1997).

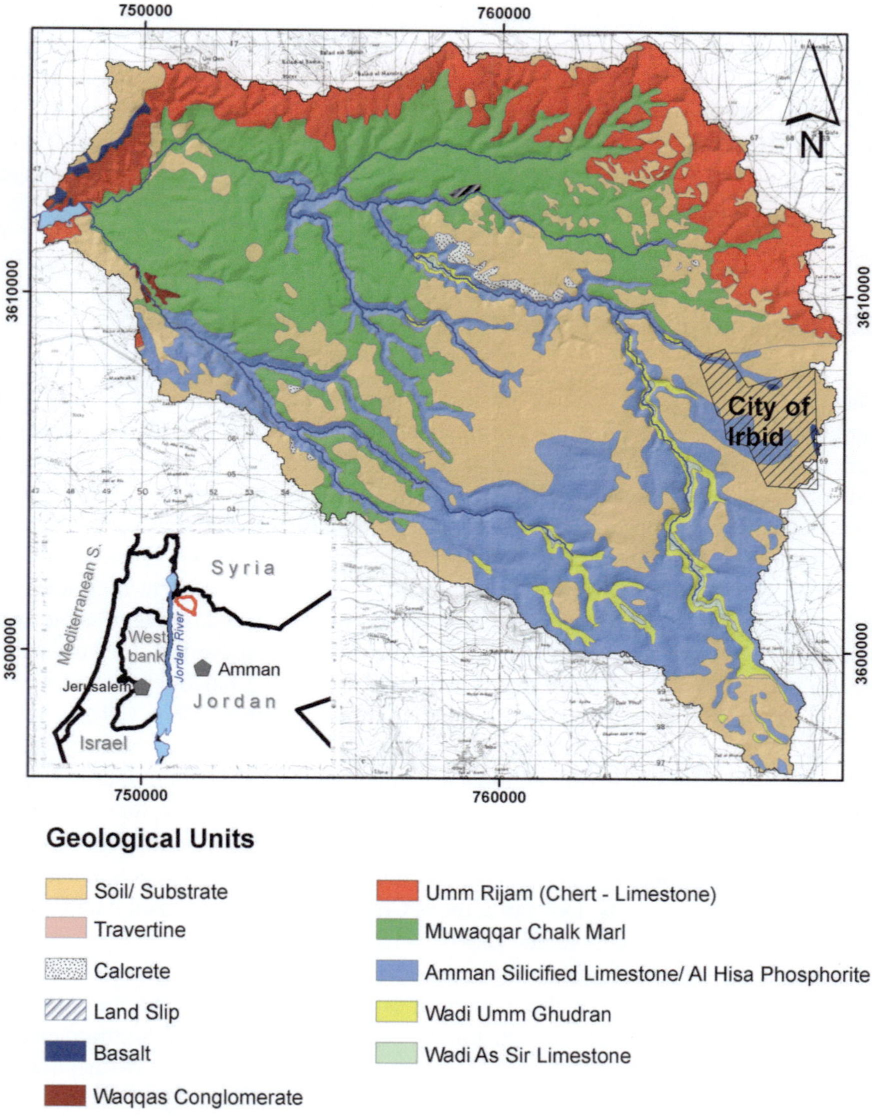

Fig. 2.5 Wadi Al-Arab geological map (MWI data 2010; Projection: UTM, WGS84, zone 36N)

An xrd analysis of the minerals conducted by Prof. Pöhlmann from the University of Halle in 2012 also indicated terrestrial input like quartz into the MCM geology, which in a shallow lagoon could be of synsedimentary as well as subsequent aeolian input (own analysis; Al-Sharhan et al. 1997).

It becomes visible in Fig. 2.5 that the main Wadi and its tributaries (= ephermal streams) have cut their way through to the ASL/AHP unit and older rocks. These units are part of the A7/B2 aquifer (Appendix 2) which is the major aquifer of the region and highly fractured and karstified. It is overlain by the bituminous MCM aquitard (B3) and the locally productive URC aquifer B4 (Rödiger et al. 2014; Siebert et al. 2014). Due to the NW-wards dip of the plateau, groundwater drainage follows this direction to the Yarmouk gorge in the north.

2.3 Geomorphological Features of the Wadi Al-Arab

The catchment displays a more or less coherent plateau with rolling hills and agricultural plains in the east. To the west and south, the relief energy is higher and agricultural areas are limited to top, saddle and footslope positions. The drainage network is assumed to be mainly controlled by faults, grabens, and lithological contact zones as it corresponds to the main fault direction (NE to SW) in the catchment. The Wadi itself is a V-shaped valley with steep flanks. The majority of the Wadi bed is only 1–3 m wide and eroded down to the basement rock leaving only shallow (<0.2 m) sediments in areas where a slight meandering is possible and small areas of point bars and cut banks can be defined. On the last 7.5 km of the stream bed, the Wadi opens up to a width of 12 m and big paleo-terraces of up to 6 m in height are present and are used agriculturally (Sect. 6.7, Fig. 6.9).

The Wadi Al-Arab is exclusively made up of carbonatic bedrock and, throughout the landscape, shows typical geomorphological features that are connected with a karstic environment such as grykes, sinkholes, uvalas, and lost rivers (Strahler and Strahler 2009). The slopes of the MCM geological unit show exposed consolidated limestone with patches of red Mediterranean soil and vegetation on the top and shoulder positions. Downhill, the MCM is present as unconsolidated marly substrate constituting steep slopes (>5–6°; Fig. 2.6).

Reasons for these differences in hardness of the same geological unit could be facies differences, as well as the hardening of the marly geology due to exposure and bio-chemical processes like crusting or cementation (personal communication Prof. Dr. Bandel, 2011). Another possible theory is adapted from the description of the "Nari" crusts in Israel (Singer 2007). These carbonatic hard layers result either from the precipitation of $CaCO_3$ from capillary waters in the rock lifted by evaporation or are the result of the leaching of carbonates in relic soils. The carbonates accumulate deeper in the soil profile or directly above the bedrock, building a cemented layer. When the soil erodes, the cemented bedrock becomes exposed (Singer 2007). During the study the genesis of these differences in the same geological unit could not be clarified finally. The ASL unit in the east also shows hard limestone in top positions with soil patches.

Fig. 2.6 Typical MCM hill with hard limestone on top and marly slopes downhill (Kraushaar 2010)

Fig. 2.7 Typical ASL hill with hard limestone along the slope (Kraushaar 2010)

On the slopes the silicified limestone leads to a more terraced morphology with a high stone content in the soil layer as seen in Fig. 2.7. A visible characteristic throughout the Wadi Al-Arab as well as in other Mediterranean regions such as reported by Kirkby et al. (1990), Cerdà (1998) and Bochet and García-Fayos (2004) is the phenotypical difference of south and north exposed slopes (Fig. 2.8).

South-facing slopes especially on marly geology contrast visibly in the density of the vegetation cover as well as the thickness of the soil layer (Sect. 2.4). Cerdà (1998) proved that for bare soils on south-facing slopes, higher bulk density values, less soil depth, and crusting are responsible for enhanced runoff and erosion processes and, for Spain, reports higher erosion than on south-facing slopes. This is due to the insufficient protective vegetation cover and hence, less organic content in the top soil layer (first 2 cm; Cerdà 1998; Bochet and García-Fayos 2004). The different erosion features and forms are described throughout Chap. 3.

Fig. 2.8 Exposition differences in Vegetation in a MCM valley. *Left* south facing slope; *Right* north facing slope (Kraushaar 2011)

2.4 Soils in Wadi Al-Arab

Many monographies have been written on the subject of Mediterranean soils and especially on red Mediterranean soils and their theory of genesis (Federoff and Courty 2013). This chapter will give a brief overview of pedogenetic processes and soil types according to the IUSS Working Group WRB (2007) that are relevant in Wadi Al-Arab and Jordan's soil mapping efforts and field observations in the research area. The aim is to provide a general overview of the pedological processes and common soil types in the research area and should be in no case judged as an attempt for a holistic overview on the subject of soils in the Mediterranean area, genesis theories or a refurbishment of the nomenclature. Thus, this chapter will focus on the IUSS Working Group WRB (2007) soil type labels because these are internationally acknowledged.

2.4.1 Mediterranean Soils and Typical Pedogenetic Processes

Wadi Al-Arab shows the characteristics of a Mediterranean region as described by Yaalon (1997) with winter rains, carbonatic bedrock material, and a relief of rolling hills with steep slopes. The winter months are more important for the soil development than the longer dry season because of the excess rain. In summer the soils and vegetation dry out (Yaalon 1997). These conditions favour certain pedological processes which lead to the typical standard winter rain Mediterranean catena with chromic Cambi-/Luvi- or Vertisols in geomorphological stable positions as plateaus, saddle, footslopes, and terraces on carbonatic bedrock. Less geomorphological stable positions on slopes are characterised by shallow or initial soil horizons above consolidated or unconsolidated marly carbonatic bedrock material, classified as Lepto- or Regosols, with a high stone content.

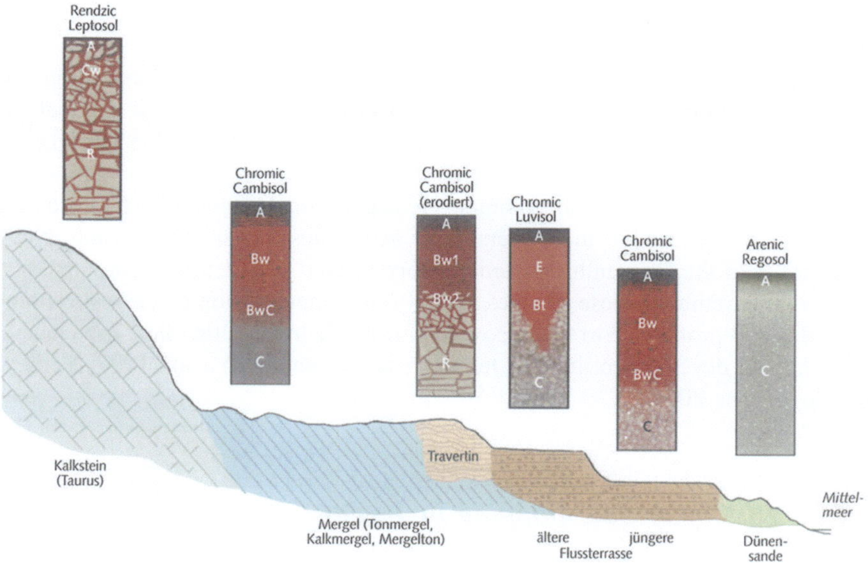

Fig. 2.9 Standard winter rain Mediterranean Catena (Zech and Hintermeier-Erhard 2002)

Lepto- or Regosols are described on slopes for coastal sand dunes by Zech and Hintermeier-Erhard (2002, Fig. 2.9), but are also reported for the region to be extensive in eroding land (IUSS Working Group WRB 2007; Verheye and de la Rosa 2009).

The genesis of the chromic soils is strongly discussed in literature and three possible coexistent processes could contribute to the formation:

(1) Solution weathering of the bedrock leaving the bedrock residue as soil matrix. The dissolution by meteoric water will eventually leach carbonates and other solubles from the soil matrix and cause the lowering of the pH, enhance the chemical weathering of the matrix leftovers to a very clayey substrate, and, finally, the migration of clay in the profile (Blume et al. 2002). In a more semi-arid climate carbonates might not be leached completely but precipitate in the lower parts of the profile as nodules, pores or ped coatings, or calcaric horizons (Yaalon 1997; Verheye and de la Rosa 2009).
(2) Allochthonous dust inputs, often responsible for the secondary liming of the soil profiles (Federoff and Courty 2013; Yaalon 1997; Herrmann et al. 2010).
(3) Isovolumetric replacement, where authigenic clay minerals grow in an over-saturated gel (Lucke et al. 2014).

The characteristic red color develops in hot and dry summers when the soil desiccates. Then iron oxyhydroxides released through weathering from the bedrock precipitate to red poorly crystalline ferrihydrates and fine-grained hematite, which coat the particles. This process, also known as rubification, can lead to very red soils with a hue redder than 5YR and a chroma of above 5 (Yaalon 1997; Verheye and de la Rosa 2009).

High proportions of swellable clays in a soil can lead to peloturbation and cracking resulting in a constant internal turnover processes of the soil material and the respective micro relief structures as the gilgai relief and visible slickensides (=Vertisols; IUSS Working Group WRB 2007).

Often the clayey soils on common cretaceous marls contain a high sodium saturation which leads to the dispersion of aggregates (Faust and Schmidt 2009). Together with the long anthropogenic history of land use, deforestation, and vegetation cover removal, these changes strongly supported erosion processes resulting in eroded soil profiles up to the uncovering of the bedrock. Other land use changes affecting soil development include terracing and irrigation (Yaalon 1997; Verheye and de la Rosa 2009).

2.4.2 History of Soil Mapping in Jordan

Jordan was the first to object to soil mapping in 1950 using the US Soil Taxonomy and identifying 12 different soil groups on a regional scale of 1:1.000.000. For areas with an average precipitation of more than 250 mm/year, as is the case in Wadi Al-Arab, mostly Verti- and Cambisol soils were identified. However, structured mapping did not start before 1989 with an international team of experts resulting in 3 levels of land and soil mapping (Table 2.1; MoA 1993; Al-Qudah 2001). Wadi Al-Arab and the region around Irbid have been subject to Level 1 and partly Level 2 mapping. This resulted in a generalized soil map with 4 different soil groups following the FAO guideline from 1979 (Fig. 2.10).

The classification groups chosen are not suitable to make more detailed statements about the relationship between relief, geology, and soil development. But they do give a broad overview of soil types that can be found in the region with the tendency of vertic Cambi- and Vertisols in more geomorphological stable relief

Table 2.1 3 Levels of soil mapping efforts in Jordan (Al-Qudah 2001)

Level	Density of observation site	Materials used	Map scale
1—General survey	7.6 km^2	LANDSAT, aerial Photography, field observations	1:250.000
2—Areas suitable for agriculture	3.5 km^2	Panchromatic SPOT imagery combined with LANDSAT thematic mapper	1:50.000
3—Priority areas	15 observation site km^{-2}	Additional field surveys	1:10.000

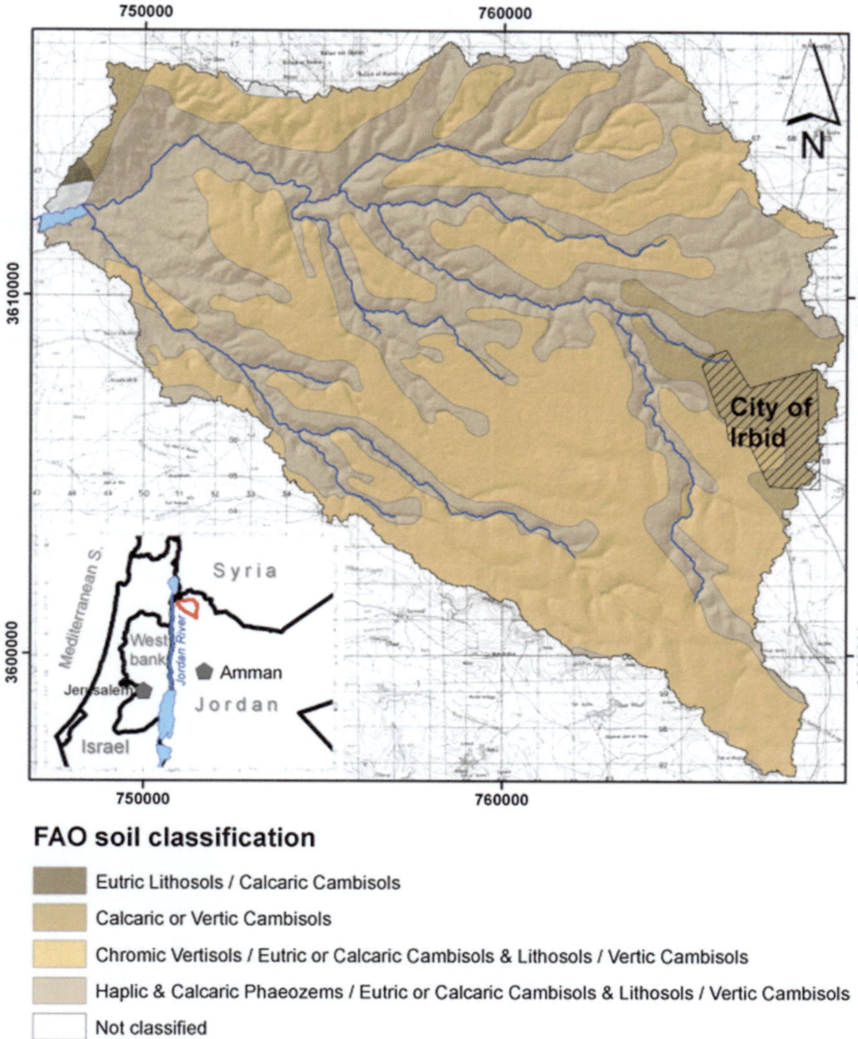

Fig. 2.10 FAO soil classification in Wadi Al-Arab (MoA 1993; Projection: UTM, WGS84, zone 36 N)

positions and calcaric Cambi- and Lithosols (Leptosol in the WRB 2007) on the slopes.

As reported by Ziadat et al. (2010), Catenal studies relating topography to soil development have rarely been done in Jordan. In Wadi Al-Arab itself, Khresat et al. (1998) as well as Lucke (2007) analysed soil profiles on different source rocks but not with the intent of mapping the area on the basis of a topographic and hence, morphologic relation.

2.4.3 Soils in Wadi Al-Arab—Own Observations

The sediment transport model demanded a soil map with various parameters for the top and subsoil horizons of the different soil types. Therefore, 28 profiles along topo sequences in the major geological units ASL/AHP, MCM, URC, and basalt were recorded and analysed. Land units characteristic of the analysed soil types were defined using morphological indicators, exposition and land use. A detailed description of the setup of the soil map of Wadi Al-Arab is to be found in Sect. 6.4 and Appendix 6.

An exemplary Catena from the soil mapping campaign is given in Fig. 2.11 for a typical valley in the north of the catchment with MCM on the slopes and ASL/AHP in the Wadi bed. Chromic Vertisols and Luvisols with a depth of more than 2 m were observed on geomorphological stable positions or in colluvial depressions (own observations; Khresat et al. 1998). Deepest soils exist on the plateau area to the east of the catchment and on the basaltic plateau in the north west (own observations, Lucke 2007). The soil development on the slopes varies with the geological parent material and its hardness. On consolidated geology (URC, MCM and ASL) usually only brown soil patches are left (Leptosols). On marly MCM slopes either Cambi- or Regosols were indentified, mainly depending on the exposition and how advanced erosion processes have denudated the soil profile (Fig. 2.11). Recent Wadi deposits are classified as Regosols, old terraces as Fluvisols.

A general compilation of the main observed characteristic of the soil types for the three major geological units are listed in Table 2.2, whereas Muwaqqar Chalk Marl (MCM) and Umm Rijam Chert (URC) tend to show similar developments on

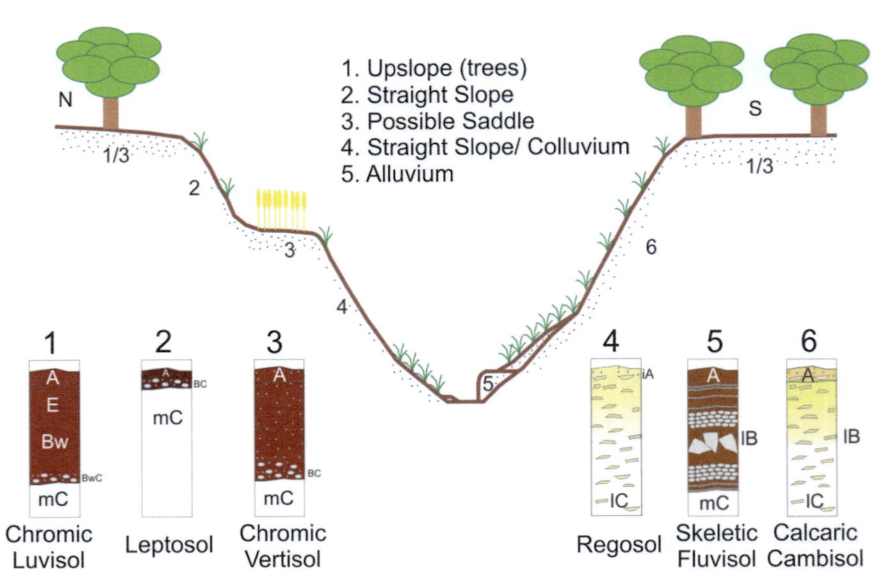

Fig. 2.11 Typical soil types in a valley with MCM on top and ASL/AHP in the valley bottom

Table 2.2 Field observations on soils in Wadi Al-Arab

Position	Muwaqqar Chalk Marl and Umm Rijam Chert	Amman silicified limestone/Al Hisa Phosphorites
Geomorphological stable position as plateaus, saddles, footslopes and terraces or soil patches on hard limestone slopes	Sediments with a more red color (5YR 3/4 to 7.5YR 2.5/4)	Sediments tend to be more brownish (10YR 4/3 to 7.5YR 4/4)
	– Sediments are very clayey (around 45 %) – Cracks and signs of pedoturbation (incorporated calcaric stones from the top, slickenslides) are often visible throughout the profile – Argillic horizons occur sometimes together with the leaching of carbonates – Sediments are mostly calcaric (c2-c4; AG Boden 2005), with visible nodules, calcareous horizons and pseudomicelia	
Slopes	– On marly substrate hardly any visible differentiation between soil horizon and geology possible – Light colors as 10YR 6/4 prevail – All sediments are carbonatic (c3-c4; AG Boden 2005)	– Shallow Ah horizon with 10YR 4/3 in color – Following horizons much yellower and extremely stony with a high silicified limestone content – All sediments are carbonatic (c3-c4; AG Boden 2005)
Wadi bed	– Wadi beds are mostly eroded to the bedrock (ASL/AHP; Sect. 2.2) – Old Wadi terraces show deep profiles with +6 m, an alternation between gravels and sands and a red soil color (7.5 4/4) – Recent Wadi deposits consist of greyish sands – Carbonatic (c4; AG Boden 2005)	

respective positions. None of the sediments examined show red colors of 5YR and a chroma >5 as often described for soils in the Mediterranean (Yaalon 1997). Low organic matter content in the A-horizons (<1 %) were analysed. Intense biological perturbation of the first 10–20 cm of the profiles was visible in the field.

A basalt profile revealed a carbonated enriched (c4; AG Boden 2005) red loam with nearly no stone content (<3 %) and a thick carbonatic layer above the underlying basalt rock. Together with the carbonatic nodules and pseudomicelia in the overlaying red-brown sediments, it indicates secondary carbonatization of aeolian character. However, in recent profiles at the Golan Heights, Jahn (1995) found only 2–5 % of the soil originating from dust input, whereas a Paleosol from the same region showed 20–40 % of clay minerals from aeolian input, postulating that this might be the case for the whole region.

In conclusion it can be stated that the general pedogenetic processes of the Mediterranean region listed in this Chapter are to be found in Wadi Al-Arab soils, too. Soil types vary more with the hardness of the geological bedrock particularly on slopes and the exposition of the slopes. Geomorphological stable positions are characterized by red or dark brown sediments. All these observations pose many additional questions to the soil's genesis, age, and environmental conditions at that time, which cannot be clarified here and are not the focus of this work. For further readings on the region please refer to Verheye (1973), Torrent (2004), Cordova et al. (2005), Lucke (2007), Lucke et al. (2014), and others.

2.5 Vegetation in Wadi Al-Arab

2.5.1 Potential Natural Vegetation of Northern Jordan

Wadi Al-Arab is bio climatically classified as Mediterranean to semi-arid following the "quotient pluviothemique of Emberger" (Al-Eisawi 1983; cited in Al-Eisawi 1996).

Covering the mountainous region of the Jordan Valley Escarpment, Al-Eisawis (1996) concludes that these characteristics support the most fertile soils and "best vegetation especially the forest climax" in Jordan (1996). Trees such as Aleppo Pine (*Pinus halepensis*), Palestinian Oak (*Quercus calliprinus and the subspecies Q. coccifera*), Valonia Oak (*Q. ithaburensis*), Carob tree (*Ceratonia siliqua*), *and* Pistacio (*Pistacia* spp.) are to be found (Lucke 2007; Al-Eisawi 1996).

The deciduous Valonian Oak forests and associated plants, like the Spiny Broom (Calycotome villosa) and the Persian Cyclamen (*Cyclamen persicum*), belong to the potential natural vegetation in the region (Al Eisawi 1996), which is still to be found mainly on top positions in soil patches on hard limestone and in areas of sparse density.

Dating back as early as the 2nd century AD the Romans prohibited lumbering, and today the government safeguards the trees through the introduction of rangers and penalties (personal communication Dr. H. Kirsch; BGR, Jordan Rangers, 2012).

2.5.2 Land Use in Wadi Al-Arab

From GLS 2010 Landsat ETM+ scenes Mallast (2011) generated a land use map for the Wadi Al-Arab showing the current most prominent landscape units (Fig. 2.12; Mallast 2011; modified Kraushaar 2012). These are by coverage percent bare soil/rock (34 %), olive orchards (27 %), areas with natural bush vegetation (=low Maquis; 11 %), agricultural fields and urban settlements each with 14 %. Trees and water together cover an area of around 0.4 %. Agricultural fields are mainly present in the Irbid basin west of the City of Irbid. In the mountainous regions NW of the catchment agricultural areas are limited to top, saddle, and footslope positions, as well as some terraces. Most of the agricultural land is reforested with rain-fed olive

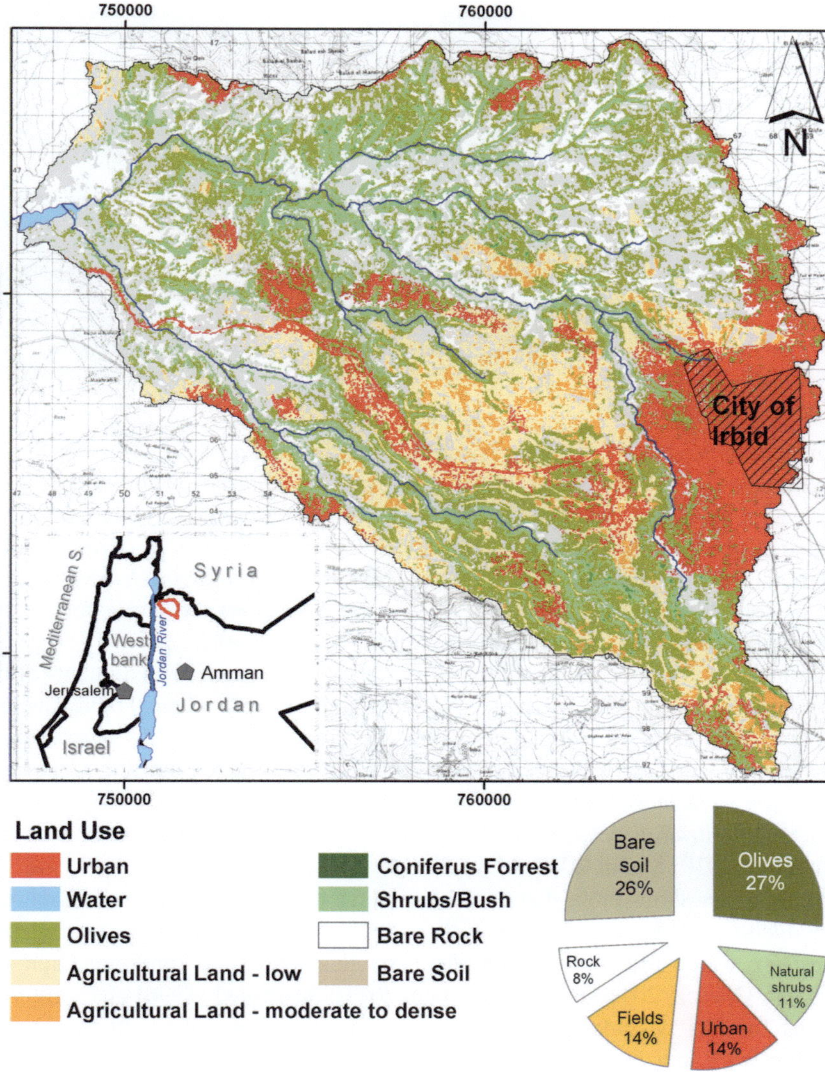

Fig. 2.12 Land use map Wadi Al Arab (Mallast 2011; modified Kraushaar 2012; Projection: UTM, WGS84, zone 36N)

trees (*olea europaea cultivar Nabali or Baladi*) of different age (personal communication with farmers, Fig. 2.12). Especially since the nineties the Government has supported olive cultivation and hence the orchards extend onto steep slopes throughout the catchment (Venot 2003; own data, 2012).

The typical most abundant landscape unit "bare soil/rock" features patchy vegetation (Fig. 2.13). Common flora found in these regions are the Spanish Nut (*Gynandriris sisyrinchium*), Common Asphodel (*Asphodelus aestivus*), Poppy

Fig. 2.13 Predominant landscape unit with common Asphodel (Kraushaar 2011)

(*Papaver Genus*), many different species of the *Asteraceae family like the* Spring Groundsel (*Senecio Vernalis*), the Cornflower (*Centaurea hyalolepis*), *Dwarf Chicory* (*cichorium pumillum*), Globe-(*Echinops Genus*), Syrian-(*Notobasis Syriaca*) and Palestine-Thistles as well as the Common Fig (*Ficus carica*), which prefers to grow in carst holes or abandoned cisterns. Sage (*Salvia Genus*), Mallow (*Malva* spp. *Malvaceae*) and Zaarta (*Origanum Syriacum*) are just a few of the many edible plants that are regularly harvested by the people.

On more marly slopes with natural shrubs and grasses different exponents of the Brassicaceae family like Mustard (*Brassica and Sinapis Genus*) and Buckler Mustard (*Biscutella didyma*), the Common Horse Shoe Vetch (*Hippocrepis unisiliquosa*), Jerusalem Buttercup (*Ranunculus millefolius*), Squirting Cucumber (*Ecballium elaterium*), Caper (*Capparis Genus*), Dwarf Mesquite (*Prosopis farcta*), Colocynth (*Citrullus colocynthis*) and Thistles are to be found (Al-Sheikh and Kumamoto 2010; Maani 2008; MWI 2011).

Many areas show biota that refers to degradation as the Common Asphodel (*Asphodelus aestivus*, Fig. 2.13) as well as the Levantine Tortoise (*Testudo graeca terrestris*), which often can be found in the field (personal communication PD Dr. habil. Ollesch, 2011 and Dr. Badarneh, 2011). Al-Eisawi (1996) gives different reasons for this degradation connected to human activity like settlement, construction work, grazing of animals, and fires etc., which are impaired by Lucke's findings (2007).

References

Abed AM, Al-Agha MR (1989) Petrography, geochemistry and origin of the NW Jordan phosphorites. J Geol Soc 146:499–506

Abed AM, Fakhouri K (1990) Role of microbial processes in the genesis of the Jordanian Upper Cretacious phosphorites. In: Notholt AJG, Jarvis I (eds) Phosphorite research and development. Geographical Society Special Publication, No 52, pp 193–203

Ad-hoc-Arbeitsgruppe (AG) Boden (2005) Bodenkundliche Kartieranleitung. Bundesanstalt für Geowissenschaften und Rohstoffe in Zusammenarbeit mit den Staatlichen Geologischen Diensten, 5th ed. Hannover

Al-Eisawi D (1996) Vegetation of Jordan. UNESCO, Cairo

Al-Qudah B (2001) Soils of Jordan. In: Zdruli P, Steduto C, Lacirignola L, Montanarella (eds) Soil resources of southern an eastern mediterraenan countries. Options Méditerranéennes, Série B, n 34, Bari, pp 127–145

Al-Sharhan AS, Nairn AEM (1997) Sedimentary basins and petroleum geology of the middle east. Elsevier Science e.V, Amsterdam

Al-Sheikh B, Kumamoto M (2010) Field guide to wild flowers of the Shahabil bin Hassneh EcoPark (EcoPeace-FoEME)

Bender F (1968) Geologie von Jordanien, vol 7. Gebrüder Bornträger, Berlin/Stuttgart

Blume H-P, Brümmer GW, Horn R, Kandeler E, Kögel-Knabner I, Kretzschmar R, Stahr K, Wilke B-M (2002) Scheffer/Schachtschabel: Lehrbuch der Bodenkunde, 15th edn. Spektrum Akademischer Verlag GmbH, Berlin

Bochet E, García-Fayos P (2004) Factors controlling vegetation establishment and water erosion on motorway slopes in Valencia, Spain. Restor Ecol 12(2):166–174

Bossone A, El-Akkad D, Sultan Z (2013) The nuclear players in the Arab world. Nature Middle East. doi:10.1038/nmiddleeast.2013.60

Cerdà A (1998) The influence of aspect and vegetation on seasonal changes in erosion under rainfall simulation on a clay soil in Spain. Canadian J Soil Sci 321–330

Cordova CE, Foley C, Nowell A, Bisson M (2005) Landforms, sediments, soil development, and prehistoric site settings on the Madaba-Dhiban Plateau, Jordan. Geoarchaeol Int J 20(1):29–56

Eltaif NI, Gharaibeh MA, Al-Zaitawi F, Alhamad MN (2010) Approximation of rainfall erosivity factors in North Jordan. Pedosphere 20(6):711–717

Faust D, Schmidt M (2009) Soil erosion processes and sediment fluxes in a Mediterranean marl landscape, Campiña de Cádiz, SW Spain. Zeitschrift für Geomorphologie 52:247–265

Federoff N, Courty A (2013) Revisiting the genesis of red Mediterranean soils. Turkish J Earth Sci 22:359–375

Flexer A (2001) The pre-Neogene geology of the Near East. In: Horowitz A (ed) The Jordan rift valley. CRC Press, Netherlands

Gómez JA, Sobrinho T, Giráldez JV, Fereres E (2009) Soil management effects on runoff, erosion and soil properties in an olive grove of southern Spain. Soil Till Res 102:5–13

Herrmann L, Jahn R, Maurer T (2010) Mineral dust around the Sahara—from source to sink. A review with emphasis on contributions of the German soil science community in the last twenty years. J Plant Nutr Soil Sci 173:811–821

Horowitz A (ed) (2001) The Jordan rift valley. CRC Press, Netherlands

IUSS Working Group WRB (2007) World reference base for soil resources 2006, first update 2007. World Soil Resources Reports No. 103. FAO, Rome

Jahn R (1995) Ausmaß äolischer Einträge in circumsaharische Böden und ihre Auswirkungen auf Bodenentwicklung und Standortseigenschaften. Hohenheimer Bodenkundliche Hefte 23

Khresat SA, Rawajfih Z, Mohammad M (1998) Land degradation in north-western Jordan: causes and processes. J Arid Environ 39:623–629

Kirkby MJ, Atkinson K, Lockwood J (1990) Aspect, vegetation cover and erosion on semi-arid hillslopes. In: Thornes JB (ed) Vegetation and erosion. Processes and environments. British Geomorphological Research Group Symposia Series. Wiley, pp 25–39

Lucke B (2007) Demise of the Decapolis. Past and present desertification in the context of soil development, land use and climate. Dissertation at the Brandenburg Technical University of Cottbus

Lucke B, Kemnitz H, Bäumler R, Schmidt M (2014) Red Mediterranean soils in Jordan: new insights in their origin, genesis, and role as environmental archives. Catena 112:4–24

Maani J (2008) Field guide to Jordan, 2nd edn. Maani Publishing, Jordan

Mallast U (2011) Klassifikationsbericht des Smart Catchments (See Genezareth – unteres Jordantal – Totes Meer) basierend auf GLS 2010 Landsat ETM + Szenen. [unpublished]

Ministry of Agriculture (MoA) (1993) National soil map and land use Project. Level 1. Hunting Technical Services LTD and Soil Survey and Land Reseach Center. Amman, Jordan

Ministry of Water and Irrigation (MWI), Jordan Valley Authority (JVA) (2011) Wild Plants in Jordan valley. Water Information and Public Awareness Department, Jordan

Moh'd BK (2000) The geology of Irbid and Ash Shuna Ash Shamaliyya (WAQQAS), Map Sheets No. 3154-II and 3154-III. Geological Directorate. Geological Mapping Division, Bulletin 46, Amman

Moh'd GK, Powell JH (2010) Uranium distribution in the Upper Cretaceous-Tertiary Belqa Group, Yarmouk Valley, northwest Jordan. Jordan J Earth Environ Sci 3(1):49–63

Peel MC, Finlayson BL, McMahon TA (2007) Updated world map of the Köppen-Geiger climate classification. Hydrol Earth Syst Sci 11:1633–1644

Renard KG, Foster GA, Weesies DK, McCool DK, Yoder DC (1997) Predicting soil erosion by water: a guide to conservation planning with the revised universal soil loss equation. Agriculture Handbook 703, United States Department of Agriculture, Washington, DC

Rödiger T, Geyer S, Mallast U, Merz R, Krause P, Fischer C, Siebert C (2014) Multi-response calibration of a conceptual hydrological model in the semiarid catchment of Wadi al Arab, Jordan. J Hydrol 509:193–206

Siebert C (2005) Saisonale chemische Variationen des See Genezareth, seiner Zuflüsse und deren Ursachen. Dissertation. Fachbereich Geowissenschaften der Freien Universität, Berlin

Siebert C, Möller P, Rödiger T, Geyer S, Kraushaar S, Dulski P, Guttmann J, Subah A (2014) Thermal waters in the Lower Yarmouk Gorge and their relation to surrounding aquifers. J Geochem [unpublished]

Singer A (2007) The soils of Israel. Springer, Berlin

Strahler AH, Strahler AN (2009) Physische Geographie, 4th edn. UTB, Stuttgart

Torrent J (2004) Mediterranean soils. In: Hillel D (ed) Encyclopedia of soils in the environment. Elsevier Academic Press, Amsterdam, pp 418–427

United Nations Environment Programme (UNEP) (1992) World atlas of desertification, vol 3/4. London

Venot JP (2003) Volume IV: the peculiar case of the olive trees orchard. Reclamation's history of the Jordan River Basin in Jordan, a focus on agriculture: past trends, actual farming systems and future prospective. INA P-G Paris-Grignon National Institute of Agronomy. French Regional Mission for Water and Agriculture (MREA) French Embassy, International Water Management Institute (IWMI)

Verheye WH (1973) Formation, classification and land evaluation of soils in mediterranean areas with special reference to Southern Lebnon. Dissertation. University Gent

Verheye WH, de la Rosa D (2009) Mediterranean soils. In: Verheye WH (ed) Land use, land cover and soil sciences, vol 8. Encyclopedia of Life Support Systems (EOLSS)

Yaalon DH (1997) Soils in the Mediterranean region: what makes them different? Catena 28:157–169

Zech W, Hintermeier-Erhard G (2002) Böden der Welt. Spektrum Akademischer Verlag GmbH, Berlin

Ziadat FM, Taimeh AY, Hattar BI (2010) Variation of soil physical properties and moisture content along Toposequences in the arid to semiarid area. Arid Land Res Manag 24:81–97

Zielhofer C, Faust D (2002) Mediterranes Klima und Mensch-Umwelt-Beziehungen. Praxis Geographie 3:10–14

Chapter 3
Erosion Features and Measurement Methods on the Slope

During field work, different erosion features were identified on different morphological positions and under different land use. On top and saddle positions fields are agriculturally used. From time to time ephermal gullies as well as piping were observed in these land units. However, most of the small-scale erosional features are destroyed quickly on fields by the regular ploughing. Under olive orchards on slopes, splash, rill and gully erosion occur and follow the downhill furrows, canalizing the runoff. Stone pedestals and siltation features in the rills and gullies are also visible and tillage erosion likely as reported by Barnevald et al. (2009) and Vanwalleghem et al. (2010). On slopes with natural shrubs and bush vegetation, signs of intensive grazing can be observed. The animals mechanically translocate the stony mulch layer on the slope, which causes a typical pattern of rhombus shaped pathways. They exists on the majority of the slopes and leave the slope with a compacted soil surface, reduced infiltration, increased runoff, a thinning of the vegetative cover and thus, reduced resistance to water erosion (Yassoglou and Kosmas 2000). Additionally, the marly soils have a high sodium content, which leads to unstable aggregates and surface crusting, delivering good conditions for runoff generation (Yassoglou and Kosmas 2000; Thornes 2005; García-Ruiz 2010, own observations). In areas with sparse or no vegetation cover the accumulation of rills, stone pedestals, gullies and badlands, as well as laminar erosion up to the point where the bedrock is exposed are visible.

These observations led to the identification of areas with certain relief and land use, which are referred to as land use unit in the course of this work. Four major land use units where defined, corresponding to the major land use units as described in Sect. 2.5. They include, nearly levelled positions used for agriculture, olive orchards on steep marly slopes (>10 %), marly slopes under sparse or no vegetation cover, and slopes with hard limestone soil patches, both the latter used for grazing.

S. Kraushaar, *Soil Erosion and Sediment Flux in Northern Jordan*,
Springer Theses, DOI 10.1007/978-3-319-31888-2_3

In all of these land use units measurements were conducted. Methods included the OSL dating of deposited sediments in two roman cisterns adjacent to fields together with catchment measurements and volume estimations (Sect. 3.1). Furthermore, the measurement of topographic olive mounds within 7 orchards (each around 800 m^2) was implemented in synergy with tree-coring and age estimation of the orchards (Sect. 3.2). On marly slopes under low Maquis, 6 erosion pin fields (each 200 m^2) were installed and on hard limestone slopes with soil patches sediment traps were built (Fig. 2.12; Sects. 3.3 and 3.4).

The methods cover different time scales: from 560 years for the fields, an average of 32 years for the olive orchards, up to two rainy seasons for the erosion pin fields, and one rain event captured by the sediment traps. Compared to erosion plot measurements, they are non- to minimally invasive without any border effects by installation and they cover all possible erosion processes. Especially the analyses of the cistern sediments as well as the olive mound method deliver long-term average erosion estimates. This carries the advantage, that in the short time frame of a PhD thesis, results do not underlie coincidental seasonal fluctuations and thus, deliver more stable comparable results.

The following chapters describe the implementation of the methods and discuss the results (Fig. 3.1).

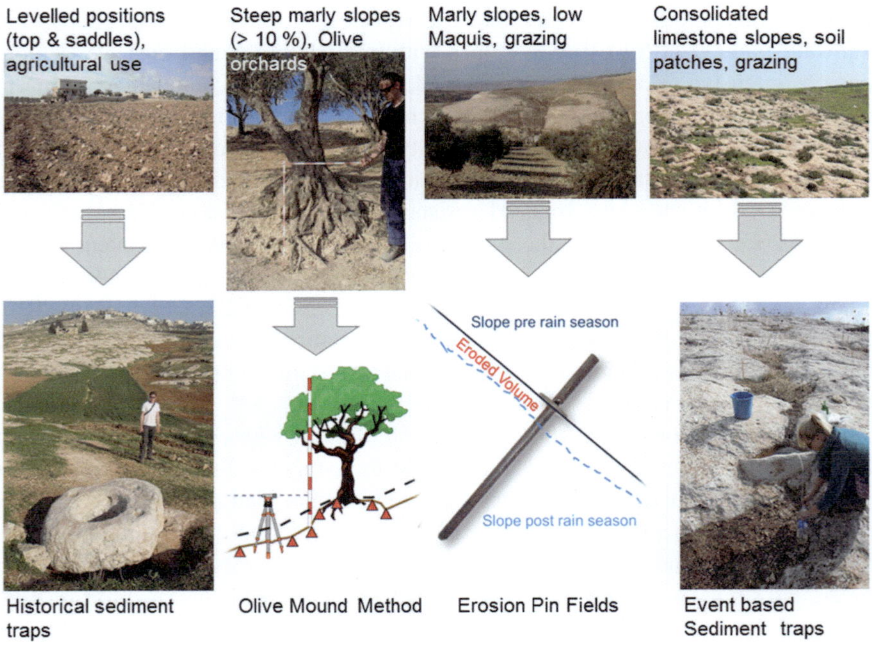

Fig. 3.1 Land use units that show erosion features and chosen measurement methods

3.1 Long Term Soil Erosion Estimates Using Roman Cisterns as Sediment Traps

The information in this Section are published by Kraushaar et al. (2015) *in Geoarchaeology 30, 369–378.*

3.1.1 Introduction

Jordan's rich archaeological record traces human occupation and settlement back to the Palaeolithic period. Northern Jordan contains a particularly rich archaeological record dating to the period when the area belonged to the eastern frontier of the Roman and later Byzantine Empires. Archaeological evidence indicates that this period (64 B.C. to A.D. 636) was the most prosperous era of the region (Mittmann 1970). Today, many former settlements can be identified by the abundance of Roman cisterns, which are sealed caverns that were chiseled into bedrock for rain-water harvesting. Most of these cisterns are today abandoned. Where the cisterns are not damaged, and their catchments have not been significantly modified by human activities, they serve as sediment traps and thus provide a historical stratigraphic archive. In small catchments where intermediate storage on hill slopes is minimal, the sediment export rate approximates the erosion rate. Although the possibility of intermediate sediment storage is minimal it cannot be entirely excluded. Thus, the volume of sediment contained within the relict cisterns in northern Jordan is used to estimate sediment yields from their individual catchments.

In addition to allowing calculation of average sediment export rates from defined catchments, these cisterns are also valuable archives of land use changes since they were abandoned. If there is high sediment connectivity from the catchments to the cisterns, the age of the basal sediments should correspond closely to the time of abandonment. Deposition within the cisterns creates sedimentary layers removed from light, allowing for optically stimulated luminescence (OSL) dating and evaluation of sediment export rates (t ha^{-1} a^{-1}) from the cisterns' catchments. Organic material contained within cistern deposits can be ^{14}C dated and provide supplemental age information. Sediment age and volume within the cistern can be related to catchment area to calculate long-term average sediment yield rates, particularly for relatively small (10–200 m^2) catchments where sediment storage is minimal and measurements are scarce. Furthermore, the method includes all major erosion processes, such as water and tillage erosion.

Over 36 archaeological sites with more than one cistern were recorded in the Wadi Al-Arab basin of northern Jordan apart from the archaeological excavations at Gadara, a city known today as Umm Qais (Fig. 3.2). In Gadara alone, 112 Roman cisterns were documented in the area (Keilholz 2007, 2012). The abundance of cisterns in the region provides an opportunity for analyzing historic erosion rates.

Here, we present stratigraphic and chronometric results from two Roman cisterns in the Wadi Al-Arab basin (Fig. 3.2). The two cisterns were selected as most suitable

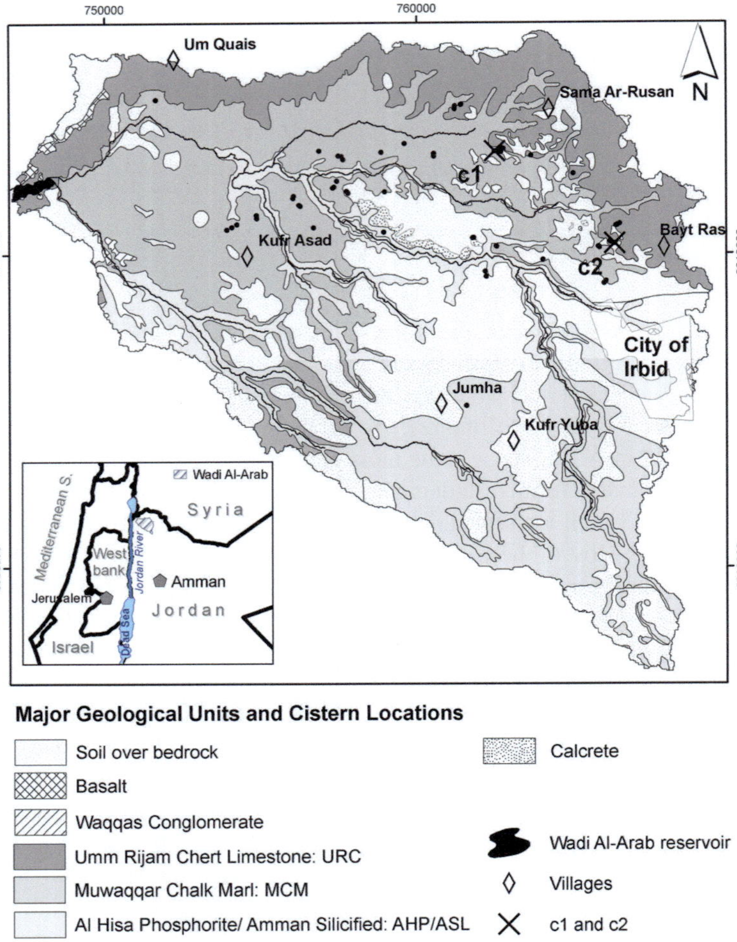

Fig. 3.2 Geology of the Wadi Al-Arab catchment with the revisited survey points of Mittmann (1970). Cretaceous and Paleocene/Eocene carbonate formations are distinguished in gray

for soil erosion studies during an interdisciplinary archaeological–geographical survey in 2010. This pilot study demonstrates the ability to date cistern stratigraphy with OSL and ^{14}C. The resulting stratigraphic chronology combined with measurements of cistern volume and catchment area can in turn be used to estimate sediment flux within small catchments, thus providing insight into local historical soil erosion.

3.1.2 Historical Background

The earliest settlements in the region around Gadara were founded in the middle of the 4th millennium B.C. During A.D. 64–324, the area belonged to the Roman

Empire and Gadara was part of the Decapolis—a federation of 10 cities in the region where Roman and Greek culture prospered. In the larger settlements such as ancient Gadara, population density is estimated to have reached around 400 inhabitants per hectare (Keilholz 2007, 2012). Demand for water in the region required the construction of hydraulic structures as early as the Iron Age (Porath 1984).

In Gadara, more than 100 man-made subsurface water reservoirs (i.e., cisterns) have been documented. These structures helped residents overcome water shortages during the dry summers and guaranteed the city a minimum of autarchy, as well as provided greater agricultural security (Keilholz 2012). The cisterns were fed either by runoff from anthropogenic surfaces (e.g., roofs) or from overland flow generated by winter rains falling on the common Mediterranean limestone landscape (Mays 2013). Since water in cisterns is stored cool and dark, the cisterns avoid contamination and are still used in the Mediterranean region (Klein 2007), especially in Jordan. Many of the Gadara cisterns were already installed and used during the Hellenistic period. Under Roman authority, aqueducts from far away springs were built leading to Gadara, which enabled greater socioeconomic development. Qanats were constructed beginning ca. A.D. 200 (Döring 2009, 2010). As these new sources of water reached Gadara, many cisterns were no longer needed inside the city. However, little is known about the installation time of cisterns in the countryside. The Roman mortar lining hints that they were at least maintained and continued to operate during Roman times and possibly later.

Around A.D. 750, a devastating earthquake destroyed large parts of Gadara and its hinterland, bringing an end to the city (Bührig 2008). During the following Abbasid and Ayyubid-Mamluk periods (ca. A.D. 750–1500) and especially in the late Ottoman period (19th/20th century A.D.), the area became repopulated, but the population density never reached its former size again (Jutta Häser & Dieter Vieweger, personal communication, 2012). As a consequence, many of the cisterns were left idle after the earthquake and have since filled with sediments.

3.1.3 Study Area

Wadi Al-Arab drains the Ajloun Plateau located ca. 500 m above sea level and joins the Jordan River at around −200 m below mean sea level (Fig. 3.2). The climate is Mediterranean to semiarid with an annual precipitation of ca. 380–530 mm; rainfall increases to the northeast. Local geology is dominated by marls, limestone, and dolomite of Upper Cretaceous and Paleogene age, resulting in alternating sequences of aquifers and aquicludes (Alsharhan and Nairn 1997; Moh'd 2000). Karst features (e.g., karren, dolines, and caverns) are common.

A plateau with rolling hills and agricultural plains characterizes the east whereas toward the west and south there is greater topographic relief associated with stream dissection. Here soils and agricultural areas are limited to top, saddle, and foot slope positions or to fluvial terraces close to the wadi bed. Cisterns for agricultural needs are usually located at shoulder positions on hill slopes; they have not been identified close to the wadi floor, on steep slopes, or on marly unconsolidated bedrock.

Most of the cisterns mapped in the Wadi Al-Arab basin were created in the Muwaqqar Chalk Marl (MCM; Fig. 3.2), an aquiclude that consists of near-surface, hard brittle limestone and soft marly rocks underneath. Its low permeability and relative softness are beneficial for the construction of cisterns, a pattern identified for chalky formations on the Israeli/Palestinian side of the Jordan River (Klein 2007: 187). Local cisterns display variable shapes and storage volumes (30–90 m^3) with different socket stones to close the inlet (Keilholz 2007, 2012).

The two cisterns selected for the current study, c1 and c2 (Fig. 3.2), harvest overland flow. Both are situated less than 50 m downslope from a modern wheat field in horizontally and vertically convex shoulder positions with patches of soil only existing in erosion-protected areas between areas of exposed limestone. Typical soils in these geomorphological positions are classified as Chromic Cambisols (MoA 1994) and range 7 to >200 cm in thickness. In 14 soil profiles located in similar positions, we calculated average grain sizes of clay 25 %, silt 68 %, and sand 7 %. Signs of active erosion on the agricultural fields include inter rills, rills, stone pedestals, and siltation features within the rills that occur after heavy rains. Most small-scale erosional features are destroyed quickly in fields by plowing, which contributes to tillage erosion. Ephemeral gullies as well as piping features were observed in these land units throughout the catchment. Average annual precipitation is 520 mm at both sites.

Cistern c1 is located ~4.5 km east of the village of Isara and is dug into the MCM on a north-facing, 8.1 ± 2.0° slope (Fig. 3.2). Cistern c2 is carved into the URC on a southeast-facing surface with an average catchment slope of 6.3 ± 1.7°, ~1.5 km west of the village of Bayt Ras. Cisterns c1 and c2 have catchment reliefs of 4.4 ± 1.3 and 3.5 ± 0.6 m, respectively. Both cisterns are about 4–5 m deep, and while Cistern c1 was left open, Cistern c2 was merely sealed with a socket stone (Fig. 3.3) that did not stop water and sediments from entering the feature.

Both cisterns contain a mortar lining, which makes it possible to distinguish cisterns from storage rooms or graves. Different layers of white, gray, and red mortar indicate the use of pure lime mortar, coal, or ground ceramics as additives. White mortar was used in early Roman times, coal mortar in the 2nd century A.D., and ceramic additives were common in the 3rd and 4th century A.D. (Porath 1984; Keilholz 2012). In later periods, both early types and transitional formulas of mortar were used (Porath 1984).

3.1.4 Methods

3.1.4.1 Field Analysis

During mid-2010 and early 2011, cistern locations were mapped in the Gadara area and the interior of 35 cisterns examined for potential further analysis. The applied criteria for suitability included intactness of mortar and ceiling, undisturbed

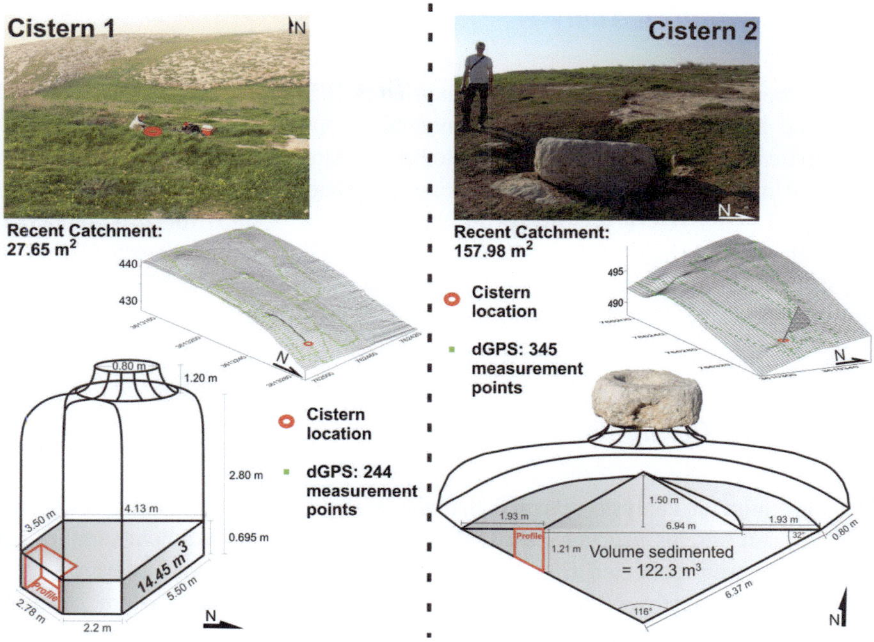

Fig. 3.3 Cisterns c1 and c2 digital elevation model with their shape and profile positions

stratigraphy, absence of water, and accessibility to work inside. Of the 35 cisterns, eight fulfilled these criteria, and the most suitable two (c1, c2) were sampled. Both cisterns contain homogenous clayrich sediments without any signs of discontinuity or bioturbation suggesting that they would be suitable for OSL dating.

Sediment profiles were characterized in the field through documentation of color, carbonate content, structure, density, and grain size, following international guidelines of the FAO (2006) and Munsell soil color chart (1994). Two to three undisturbed sediment samples for the analysis of bulk density, clay mineral composition, grain size, OSL, and gamma emitting radionuclides were taken from the upper and lower horizons of each profile. In addition, both the interior volume of the caverns and the sediment volumes of the cisterns were calculated.

Each cistern catchment was surveyed along transects using a Leica 900 dGPS with a vertical accuracy of <1 cm. From the 345 (Cistern c1) and 244 (Cistern c2) measurement points (=1.6 point m^{-2}), digital elevation models of the area were generated with ArcGIS by interpolating, where ordinary kriging (trend removal of two) yielded the smallest root mean square error. Watersheds were delimited by implementing the hydrological toolset from the program ArcGIS.

3.1.5 Laboratory Analysis

Undisturbed sediment samples were oven dried at 105 °C until constant weight and measured for their bulk density. Clay mineral composition was analyzed using a PANalytical XPert PRO X-ray diffractometer (Almelo, Netherlands) at Martin Luther University Halle. The grain size distribution was analyzed with the SEDIMAT 4-12 (UGT, Germany) at the Helmholtz Centre for Environmental Research—UFZ Halle. Gamma-emitting radionuclides were detected with a GAMMA-X, N-type coaxial HPGe detector (type GMX90-S) (ORTEC spectrometry, USA) (Tables 3.1 and 3.2).

3.1.6 OSL Dating

The clay mineral analysis revealed that the predominantly carbonatic sediments contain quartz grains necessary for OSL dating. In total five sediment samples for OSL dating were collected from the cisterns. Two samples from Cistern c1 were collected from depths of 43 and 63 cm. Three samples from Cistern c2 were collected at depths of 34, 117, and 71 cm. Sampling took place during the night using red LED headlamps (640 nm). Profiles were cleaned to remove light-exposed material, and collected sediments were placed directly into opaque plastic bags.

To determine the equivalent dose (De), the quartz fine-grain (4–11 μm) fraction was separated. After wet sieving, the sediment was then treated with HCl and H_2O_2 to remove any carbonates and organics. To obtain pure fine-grain quartz extracts, the polymineral samples were etched in 34 % H_2SiF_6 for several days (Fuchs et al. 2005). The purity of the quartz extracts was checked by infrared stimulated luminescence (IRSL) measurements and aliquots with IRSL/OSL ratios greater than 3 % were rejected (Table 3.2).

Luminescence measurements to determine the equivalent dose (D_e) were conducted on a TL/OSL-DA-15 Risø reader, equipped with blue LEDs (470 ± 30 nm) for stimulation, a Thorn-EMI 9235QA photo-multiplier combined with a 7.5 mm U-340 Hoya filter (290–370 nm) for detection and a $^{90}Sr/^{90}Y$ β-source (1.84 GBq)

Table 3.1 Sample characteristics from cistern 1 and 2

Sample_depth (cm)	Geology	Clay (%)	Silt (%)	Sand (%)	Bulk density (g/cm-3)	Water cont. (%)
c1_34	MCM	46.5	51.4	2.1	1.15	24
c1_63	MCM	n.a.	n.a.	n.a.	1.15	27
c2_43	URC	36.7	44.3	19.0	1.14	27
c2_71	URC	n.a.	n.a.	n.a.	1.12	25
c2_117	URC	n.a.	n.a.	n.a.	1.14	26

n.a. = not available

Table 3.2 OSL analytical data of cistern 1 and 2

Sample_depth (cm)	^{238}U	^{232}Th	^{40}K	$\dot{D}$ (Gy ka)	D_e	Age [ka]	Analytical uncertainty (1σ) (ka)
c1_43	7.04 ± 0.24	4.44 ± 0.25	0.64 ± 0.02	2.92 ± 0.16	1.83 ± 0.10	0.63	0.07
c2_34	7.59 ± 0.16	5.18 ± 0.25	0.68 ± 0.02	3.16 ± 0.17	3.61 ± 0.40	1.14	0.16
c2_71	7.31 ± 0.08	5.42 ± 0.25	0.93 ± 0.03	3.31 ± 0.18	4.02 ± 0.15	1.21	0.14
c2_117	6.96 ± 0.08	4.44 ± 0.25	0.69 ± 0.01	2.94 ± 0.16	3.39 ± 0.24	1.15	0.12

Note For dose rate calculation, a water content of 20 % and an a-value of 0.035 were used for irradiation. The single-aliquot regenerative (SAR) dose protocol proposed by Murray and Wintle (2000) was applied for D_e determination. Six regeneration cycles were used with shine-down curves measured for 40 s at elevated temperatures (125 °C), using a cut-heat for the test dose of 160 °C. Based on dose recovery and preheat plateau tests, a preheat temperature ranging from 220 to 260 °C was chosen for the natural and regenerated OSL signals. The results of the dose recovery and preheat plateau test indicate that the given doses can be reproduced in temperatures ranging from 220 to 260 °C.

Up to 24 aliquots per sample were measured for De determination. De calculations are the mean values of n aliquots for each sample, where n is the number of aliquots that passed the following criterion: a recycling ratio of 1 ± 0.1 and a recuperation value of 5 % (Murray and Wintle 2000). The standard error of the mean was used as De error, having considered a systematic error resulting from the β-source calibration.

The dose rates ($\dot{D}$) were obtained from sediment samples taken from within a 30 cm radius of each OSL sample location. Low-level high-resolution γ-spectrometry was applied to determine the contents of ^{238}U, ^{232}Th, and ^{40}K. Dose rate conversion factors were used from Adamiec and Aitken (1998). For all samples, an a-value of 0.035 (Mauz et al. 2006) was used to consider the α-effeciency (Table 3.2).

The cosmic-ray dose rates were calculated according to Prescott and Hutton (1994). The water content of the samples was estimated based on their porosity. Thereafter, possible minimum and maximum water contents were determined by the pore volume, and an average value of this range was chosen. The chosen values for the water content of the samples were checked by measuring the in situ water contents of each individual sample.

From sample c1_63 cm, no quartz for an OSL measurement could be extracted. However, a pottery sherd found at the depth of the sample contained a dark organic core originating either from charred traces of tempering agents used in the manufacturing of the pot or absorbed oils or other organics while heating. The sherd was crushed and pretreated with acid to remove carbonates and the bulk organic carbon was ^{14}C dated (AMS Standard) at the laboratories of Beta Analytic, Inc., in Miami, FL (Beta-327418). For calibration, the INTCAL09 data set was used, delivering three possible 2 sigma calibrated age ranges spanning 1010–1220 cal. year BP. The midpoint of the total age range and its standard deviation (1115 ± 105 cal. year BP) was used for the sediment export calculations (Table 3.3). The resulting age is an average of any organic tempering agents, any absorbed organics from the time of use of the vessel, any carbon-stained clay from the time the pot was manufactured, and any organics that were absorbed from the surrounding burial soil. We interpret most of the organic material to have come into contact with the pot during usage. Regardless, manufacturing or usage must have occurred prior to vessel destruction and deposition in Cistern c1. Thus, the age of the sherd corresponds to the maximum possible age of the sediment.

Table 3.3 Radiocarbon dating information

Sample_depth (cm)	Lab code	Measured ^{14}C Age (BP[a])	Material	$^{13}C/^{12}C$ [‰]	Calibration dataset	Conventional ^{14}C Age (BP[a])	2σ calib. age ranges (cal year BP[a])	Mid point 2σ calib. Age ± SD (cal year BP[a])
c1_63	Beta—327418	1230 ± 30	Potsherd with charring of tempering agents used in the pots manufacture, or the absorption of oils or other organics while heating	−27.7		1190 ± 30	1010–1030 1060–1180 1210–1220	1115 ± 105

[a]*BP* radiocarbon years before present, "*present*" A.D. 1950

3.1.7 Results and Interpretations

The Cistern c1 profile extends to the floor of the structure, whereas the Cistern c2 profile documents stratigraphy above the depocenter (Fig. 3.3). Both cistern profiles contain dark-brown (7.5YR 5/6) sediment. The upper 30–35 cm of cistern deposits is subangular blocky and bedded with small carbonatic rock detritus representing the youngest material. These deposits overlie a layer of massive, partly layered homogenous sediment of the same color, which in Cistern c2 is interrupted by a carbonatic stone layer within a matrix of coarse sand at 80 cm depth. Lower deposits in both cisterns contain angular blocky structure suggesting abiotic pedogenesis. Profile thickness reaches 69.5 cm (c1) and 121.0 cm (c2), and carbonate contents of 10–25 %. Cistern c1 contains a 2 cm thick ceramic and charcoal mortar layer above the bedrock contact, whereas Cistern c2 has only a single charcoal mortar layer of 1 cm.

Topographic measurements indicate catchment sizes of ~28 and ~158 m^2 for Cisterns c1 and c2, respectively. Sediment volumes were calculated in Cistern c1 assuming a cuboid (see dimensions in Fig. 3.3) and subtracting a prismatic volume given a missing corner resulting in 14.5 m^3. For Cistern c2, the cone shapes of the cistern and accumulated deposits on top were calculated following the measured and estimated dimensions (Fig. 3.3), resulting in a volume of 122.3 m^3. For the erosion calculations, only the sediment volumes from the dated horizon up to the surface (plus the sediment cone on top in Cistern c2) were used.

Sediment characteristics for the chronometric samples and their analytical results are listed in Tables 3.1 and 3.3 and Fig. 3.4. The ^{14}C age from 63 cm depth in Cistern c1 is interpreted as maximum age of the basal sediments assuming that the destruction of the ceramic vessel and sedimentation took place without any great temporary hiatus. Organics within the sherd yielded an age of 1115 ± 105 cal. year BP (=1.18 ± 1.11 ka). Sediments at 43 cm (Cistern 1) depth yielded an OSL age of 0.63 ± 0.07 ka. Deposits at 117 and 34 cm depth within Cistern c2 generated OSL ages of 1.15 ± 1.12 and 1.14 ± 0.16 ka, respectively. The uncertainty of the sherd's deposition in Cistern 1 and the position of the profile away from the center of Cistern c2 (Fig. 3.3) both prevent final conclusions regarding when the cisterns were abandoned. However, it can be postulated that Cistern c1 cannot contain sediments older than 1.18 ± 1.11 ka, and that the abandonment and first infilling of Cistern c2 must have happened prior to 1.15 ± 0.12 ka.

Sediment ages from both cisterns are compatible within their analytical uncertainties and in chronostratigraphic order. The deposit containing sample c2_71 cm shows a distinct fining-upward sequence characteristic of a single deposition event. The coarser sediments of this sequence date to 1.21 ± 0.14 ka.

Based on the four OSL and one ^{14}C ages, we estimate minimum and maximum sediment yield in tons per hectare and years (t ha^{-1} a^{-1}) for each cistern (Table 3.4). Sediment volumes in each cistern were calculated from the lower and upper horizons. Their respective sedimentation periods were assumed using their

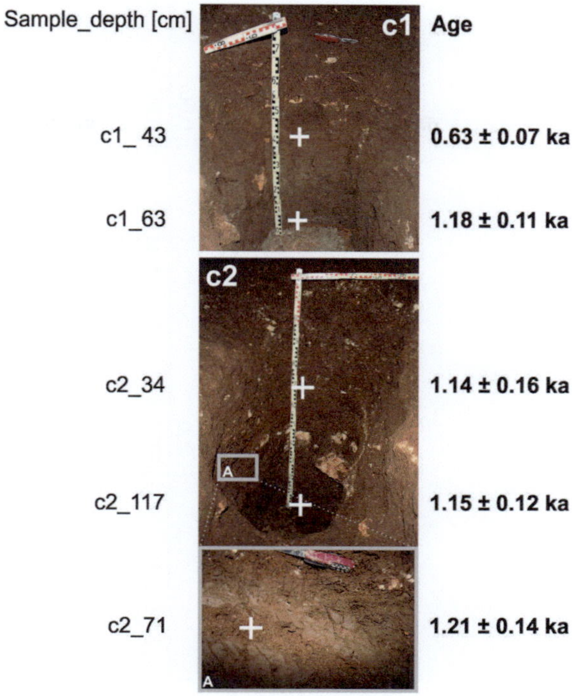

Fig. 3.4 Cisterns c1 and c2 profile stratigraphy, sample locations, and sediment ages

minimum and maximum ages until the year of measurement (2012). The results show a general mean of 4.9 ± 0.05 t ha^{-1} a^{-1} and range between 2.6 and 6.6 t ha^{-1} a^{-1}. For mass calculations, an average bulk density of 1.15 g cm^{-3} for Cistern c1 samples and a mean of 1.13 g cm^{-3} for Cistern c2 samples were used (Table 3.1); all samples have standard deviations in bulk density of 0.012 g cm^{-3}. As bulk density is the sensitive parameter for these calculations, standard deviation was used to quantify uncertainty of the sediment yield estimates (Table 3.4).

3.1.8 Discussion

3.1.8.1 Dating Reliability

The precise temporal relation of the single ^{14}C age to deposition within the cistern is undefined. In this study, we assume no significant temporal hiatus between the sherd's manufacture, use, and deposition in the cistern. This approach gives the longest possible sedimentation period and thus the lowest possible sediment yield values. However, if the sherd was manufactured and used decades before it was destroyed and incorporated into the cistern fill, the apparent age of deposition would be overestimated and the calculated sediment export values would be

Table 3.4 Parameters for the calculation of the erosion values in both cisterns

Sample and volume used for calculation	Catchment Estim.	Sediment volume (m^3)	Catchment (m^2)	Sediment yield [$t\ ha^{-1}$]	Age* (a)	Min./Max. period of sedimentation* (a)	Sediment yield [$t\ ha^{-1}\ a^{-1}$]	Uncertainty [$t\ ha^{-1}\ a^{-1}$]
c1_43 cm until surface	dGPS	8.9	27.7	3716.8	630 ± 70	560	6.64	0.07
			27.7	3716.8		700	5.31	0.06
c1_63 cm until surface	dGPS	13.1	27.7	3369.0	1177 ± 105	1072	5.08	0.05
			27.7	3369.0		1282	4.25	0.04
c2_34 cm until surface	dGPS	47.1	158.0	3946.3	1140 ± 160	980	3.44	0.04
			158.0	3946.3		1300	2.59	0.03
c2_117 cm until surface	dGPS	93.3	158.0	6673.7	1150 ± 120	1030	6.48	0.07
			158.0	6673.7		1270	5.25	0.06
						Mean	4.88	1.4

*Reference year 2012

underestimated. In contrast, OSL dating determines the age of the last episode of sediment transport and final deposition. In this study, several indicators support the correctness of the OSL ages including the following:

1. OSL characteristics such as brightness of the sample, OSL growth-curve behavior, and dose recovery tests are very good.
2. The one ^{14}C age appears compatible with OSL results from Cistern c1 within the analytical uncertainties (Fig. 3.4).
3. The OSL ages in Cistern c2 are in chronostratigraphic order.

All the numerical ages refer to the main sedimentation period after the major earthquake in *ca.* A.D. 750 and thus provide a *terminus ante quem* that corresponds with the historical narrative of the region's abandonment from various archaeological proxy data. However, the age of the sherd in Cistern c1 gives only a maximum possible age of the associated deposits, and sedimentation could have taken place later. A sediment volume of $\sim$29 m^3 was already deposited in Cistern c2 before the dated lower horizon at 1.15 $\pm$ 0.12 ka. It is not clear from the analysis if this deposition happened before or after the mentioned earthquake and subsequent abandonment.

3.1.8.2 Erosion Rates

Although there are differences in catchment size and shape of the two cisterns, both catchments are similar in slope, annual precipitation, sediments characteristics, and the assumed erosion processes that took place, that is, water erosion in the form of inter-rills, rills, and translocation by the plow. The averaged sediment export values of catchments for both cisterns cover a small range (2.6–6.6 t ha^{-1} a^{-1}) and values calculated from the depositional level and ^{14}C age of the sherd (4.3–5.1 t ha^{-1} a^{-1}) fit the range calculated from the OSL ages, thus supporting the accuracy of the results.

The deposited maximum volumes are 544,000 g m^{-2} (Cistern c1) and 667,000 g m^{-2} (Cistern c2), which equals a minimum soil cover loss of 47–59 cm from the cisterns' catchments over a maximum period of $\sim$1300 years. These calculations are based on recent catchment size measurements. However, due to changes in agricultural practices (e.g., direction of furrows) over time, catchment size might have slightly changed. Such changes are not reproducible today and could be a source of uncertainty regarding the calculations but are assumed to be of minor effect.

The calculated sediment yield rate of 2.6–6.6 t ha^{-1} a^{-1} is difficult to compare with values from other regions. Particularly in the Mediterranean, erosion and sediment yield rates are subject to remarkable uncertainty (García-Ruiz et al. 2013), since their calculation is strongly based on both the observed scale and the applied method of measurement (Fleskens and Stroosnijder 2007). Additionally, their estimations are prone to a variety of effects, including the connectivity of detached particles to small channels, intermediate storages in small catchments (<10 km^2), and periods of strong precipitation or drought (García-Ruiz et al. 2013). Hence, erosion studies in the circum-Mediterranean region reveal extremely variable export rates for agricultural fields. Kosmas et al. (1997) report 0.15–0.9 t ha^{-1} a^{-1} on cereal fields, whereas Vandaele et al. (1996) calculated 1.0–6.8 t ha^{-1} a^{-1} for fields containing ephemeral gullies as major sediment exporters. Export rates as high as 107 t ha^{-1} a^{-1} have been estimated for agricultural fields in Spain (De Santisteban et al. 2006). The sediment yields derived from our cisterns (2.6–6.6 t ha^{-1} a^{-1}) fall into the lower range of measured values from the Mediterranean but nonetheless document not insignificant soil loss. Results are coherent with recent field observations around the cisterns, where slope-parallel tillage channelizes runoff and increases rill erosion.

Age estimates for Cistern c1 suggest accelerated sedimentation through time as 4.2 m^3 sedimentation occurred between 1.18 ± 0.11 and 0.63 ± 0.07 ka and 8.9 m^3 from 0.63 ± 0.07 ka until today (Fig. 3.5). In contrast, overlapping OSL ages from Cistern c2 imply possible rapid sedimentation of 46.3 m^3 of the total 93.3 m^3 between approximately 1.1–1.2 ka. However, more age control is needed from the cistern fill to confidently quantify changes in sedimentation rate.

Based on archaeological surveys in Palestine, Gibson (2001) postulated that widespread terracing was implemented as early as Iron Age II (1000–500 B.C.), a time of stronger organizational authority, economic stability, and population pressure. During Roman times (A.D. 64–324), when the Decapolis region continued to prosper, similar circumstances led to farming activities and settlements on more or less every possible landscape position in Wadi Al-Arab (Glueck 1942; Mittmann 1970; Vieweger and Häser 2010). This leads to the assumption that existing terraces were under continuous use during the Roman period, and maintenance or possible further agricultural expansion occurred. The abandonment of the area was a result of the weaker organizational authority of the post-Roman Empire during the Abbasid and Ayyubid-Mamluk period (ca. 750–1500 C.E.). The region was less

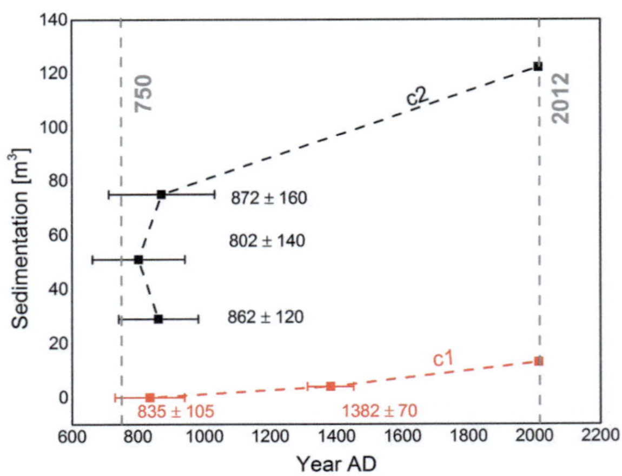

Fig. 3.5 Sediment volume versus time in Cisterns c1 and c2

resilient to environmental disturbances, such as the devastating earthquake of ca. A. D. 750. Recent land abandonments in southern Spain show that unmaintained terraces eventually collapse and yield high sediment loss (Koulouri and Giourga 2007). Thus, Wadi Al-Arab may have been witness to periods of accelerated erosion after the abandonment of the region. On the other hand, land abandonment may also result in plant succession and reduced erosion rates (Grove and Rackham 2001). Currently, we cannot determine the precise periods of pronounced erosion in the region. Further analyses of cistern sediments with higher temporal resolution are needed to address this question.

3.1.9 Conclusion

Roman cisterns in northern Jordan have served as historical sediment traps since their abandonment. Sediment profiles of two cisterns were analyzed and dated with OSL and ^{14}C, revealing a maximum first sedimentation age ca.

A.D. 760–860, which generally agrees with the archaeological evidence supporting abandonment of the region in the mid-8th century. However, as discussed, due to uncertainty of the ceramic ^{14}C date in relation to deposition within the cistern, and the position of the stratigraphic profile in Cistern c2 above the bottom of the feature, the earliest cistern stratigraphic dates only provide limiting ages for cultural abandonment.

Based on chronometric results and detailed GIS-based topographical analyses of the cistern catchments, long-term average soil loss is estimated at 2.6–6.6 t ha^{-1} a^{-1}, a range that is compatible with other soil erosion studies in the Mediterranean. These cistern sediment archives do not suffer from short-term weather fluctuations, such as can be the case during a time-limited plot study, but cover all kinds of weather events. The method presented here is minimally invasive and no edge effects occur, as instrumental installations are not needed. Hence, the use of cistern sediments to calculate long-term sediment export rates has considerable advantages compared to conventional methods used for short-term erosion observations. The abundance of Roman cisterns as mapped during this study and documented in the region provide further advantage when aiming to determine various long-term average sediment yield rates from nearly levelled positions and across regions.

Nonetheless, suitable cisterns for measuring long-term soil erosion rates have to be chosen very carefully. The cisterns should contain mortar linings confirming their use for water storage and impermeability and lack evidence of anthropogenic disruptions that might compromise stratigraphy. In our study, numerical ages were in chronostratigraphic order, but stratification could only be demonstrated for one of the two cisterns, raising important questions about the frequency of erosion and depositional events.

Despite the uncertainties, we conclude that the strati- graphic analysis and dating of cistern sediments is a useful approach for interpreting historic surficial processes and landscape change. With further implementation and method development, this approach could be a useful tool for estimating average sediment export rates from catchments in the Mediterranean region.

3.2 Mound Measurements—Quantifying Medium-Term Soil Erosion Under Olive Trees in Northern Jordan

The information in this Chapter are published by Kraushaar et al. (2014) *in Geomorphology 213, 1–12.*

3.2.1 Introduction

For the north of Jordan, Khresat et al. (1998) sketched different forms of soil degradation, mentioning that water erosion and improper farming practices are two of the major causes. Both processes aggravate each other on olive plantations in the research area, where the total area of olive trees more than doubled between 1994 and 1999 and is still increasing (Venot 2003; IOC 2005; own observations), covering around 30 % of the arable land (IOC 2005, estimation for 2006).

Many researchers have tried to demonstrate the vulnerability of olive orchards to soil erosion with the expansion of olive fields in European Mediterranean countries due to the EU Agrarian Policy and the strengthening of the national and international markets. Kosmas et al. (1997) found that if annual vegetation and plant residues are left as protective cover under the olive trees, soil loss is very low (0.008 t ha^{-1} $year^{-1}$). However, many other publications refer to olive orchards on slopes under conventional or even no tillage or cover strips to be highly vulnerable to erosion (Pastor and Castro 1995; Gómez et al. 2003, 2004, 2005, 2008a; Milgroom et al. 2005; Francia Martínez et al. 2006; Ramos et al. 2008; Barnevald et al. 2009; García-Ruiz 2010; Vanwalleghem et al. 2010). For instance, in a modeling study of olive culture on slopes up to 20 %, Gómez et al. (2003) predict water erosion to reach 80 t ha^{-1} $year^{-1}$, depending on tillage practice and the existence of cover crops. The maximum value was found for no-tilled orchards that suffer significant compaction during harvest with the beginning of the wet season (Gómez et al. 2003). Gómez et al. (2008a) found an average soil loss of 46–78 t ha^{-1} $year^{-1}$, and even in terraced systems the soil loss reached 20 t ha^{-1} $year^{-1}$. Similar results, but of lesser magnitude, were found in plot experiments. The comparison of different management systems showed 4 t ha^{-1} $year^{-1}$ eroded sediment for conventional tillage and 8.5 t ha^{-1} $year^{-1}$ for no tillage farming (Gómez et al. 2004). Francia Martínez et al. (2006) published plot results of SE Spain with a soil loss of 10.5–47.7 t ha^{-1} $year^{-1}$, corresponding to those obtained by Gómez et al. (2003, 2008a).

Using two independent tracer methods, Barnevald et al. (2009) postulate that for orchards in north-west Syria a single up and down tillage pass on a 25 % slope can cause soil loss of up to 50 t ha^{-1} $year^{-1}$. Based on repeatedly obtained high-resolution DEMs, Ramos et al. (2008) quantify possible surface reduction in certain areas of olive orchards of up to 0.2 ± 0.1 m per year with a slope ranging from 2 to 45 % and conventionally tilled.

A literature review by Fleskens and Stroosnijder (2007) on erosion below olive trees points out that the estimation of soil loss rates generally varies by four orders of magnitude. Reasons for this include the individual characteristics of an orchard, the methods used for quantifying erosion, and the spatial and temporal scales of analysis. Most of the reported erosion rates since 1990 exceed the assumed soil renewal rate by weathering, which is about 1 t ha^{-1} $year^{-1}$. Fleskens and Stroosnijder (2007) argue that the alarming erosion rates based on simple empirical models do not take into account important parameters (e.g., rock fragments protecting the soil surface or work with simplistic factors). They suggest that soil erosion rates are unlikely to exceed 10 t ha^{-1} $year^{-1}$. However, Gómez et al. (2008b) criticise Fleskens and Stroosnijder's (2007) incomplete interpretation of selective literature findings and doubt the conclusion drawn from the limited and short-term plot experiments. They believe that rates higher than the mentioned 10 t ha^{-1} $year^{-1}$ are often possible.

It becomes obvious that additional field studies are necessary to identify triggers of erosion processes and to estimate soil loss over extended time periods to reliably quantify the influence of sediment delivery from olive groves.

This is of special interest due to two regional developments:

(1) Since the beginning of the 1990s, the Jordanian government has been encouraging Jordanian farmers to grow olive trees by lowering the price of seedlings or giving them out for free (Venot 2003). Reasons for this governmental policy are to "stem rural depopulation by providing jobs for roughly 6 % of the working population" (IOC 2005, p. 2), market demands from a constantly increasing population, the cultural heritage connected to olive farming, and the fact that olive trees are ecologically adapted and easy to farm (Venot 2003, p. 4). This strategy tripled the olive production from ca. 40,000 t in 1990 to 120,000 t in 2000 and increased the orchard area significantly (Venot 2003).

(2) The EU is the main export partner for agricultural products (51.2 % in 2004) from Jordan. The extension of the EU trading zone to the Middle East and North African countries in 2014, and hence the partial liberalization of the market, will most likely lead to higher market demands in Jordan. As an example of the expected future olive oil export in the region, an increase of up to 167 % (reference year 2007) has been calculated exemplary for Tunisia (Euro-Mediterranean Agreement 2002/357/EC 2002; Jaquet et al. 2007). Similar trends may also be expected for Jordan and the olive tree-dominated Wadi Al-Arab catchment. Furthermore, a stronger intra-trade between these new adjoining countries is expected.

Consequently, more research on soil loss rates and changes in key soil parameters "is urgently required to develop effective erosion control policies" (Gómez et al. 2003). However, the average quantification of soil erosion and sediment deposition remains to be a challenge because the use of common methods such as small-scale experimental plots and models have various restrictions. Sediment plots are limited in space and time and afflicted by border effects (Fleskens and Stroosnijder 2007). Additionally, it is difficult to extrapolate the data to entire catchments because they do not represent all relevant processes at the catchment scale. Rainfall erosion events during the monitoring period might not reflect the full variability of meteorological and hydrological conditions. In addition, the estimation of soil erosion based on modeling is hampered substantially by the lack of available data. Therefore, the current study adapted a method presented by Vanwalleghem et al. (2010) that considers all acting erosion processes and delivers mean annual erosion rates considering a longer time span using the micro topography in orchards, namely olive mounds. The technique includes the measurement of the recent surfaces, the reconstruction of the historical ones using the point of germination of the trees, and the final calculation of the erosion volume. Similar attempts using residual erosion markers like pedestals of the Hawaiian silvers word to estimate soil erosion were conducted by Peréz (2007).

Using this modus operandi, the present chapter aims to answer the following scientific questions:

- Is the measurement of the olive mound a suitable method to estimate medium-term erosion rates in the region?
- What are the erosion rates under olive orchards in the study area and what are the driving factors?
- What are the characteristics of the research sites, and what implications do these have on the catchment as a whole?
- What are the possible management options to reduce erosion and increase the lifetime of the Wadi Al-Arab reservoir?

3.2.2 Study Area

In Wadi Al-Arab, around 27 % of the area is presently used for agriculture, of which 84 % is reforested with rain-fed *Olea europaea L.* cultivar Nabali or Baladi of varying ages (Mallast 2011, personal communication with farmers, 2010).

The majority of the olive orchards in the region are shared family heritages, and most of them are not accessible by tractor and tilled only by a single-furrow donkey-drawn metal plough. During the last few years, sloping areas in the unconsolidated Muwaqqar Chalk bedrock near the dam have been widely cultivated with orchards, allowing the usage of heavy machinery.

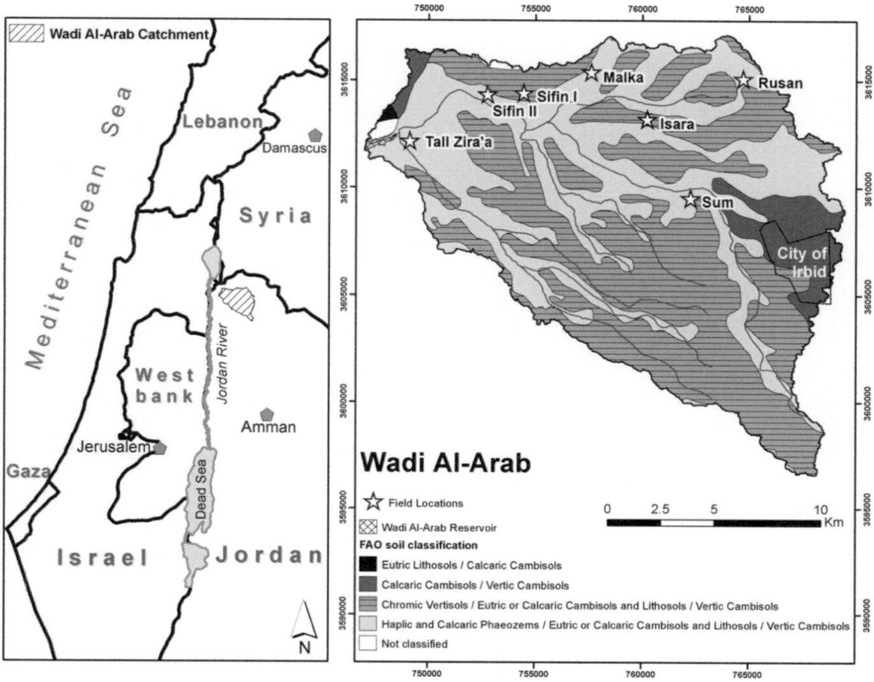

Fig. 3.6 Wadi Al Arab catchment and location of the seven research fields (projection: UTM, WGS84, zone 36 N)

To apply the mound measurement method, seven olive fields (Fig. 3.6) with clearly visible olive mounds and/or exposed roots were chosen for measurement. Selected orchards are located on the regional typical soil types: (i) Vertic Cambisols, (ii) Chromic Vertisols (upslope), and (iii) Calcaric Lithosols and Cambisols (MoA 1994). Average slopes range from 5° to 16° where olive mounds are clearly developed and can be distinguished from earth banks caused by ploughing. Rock fragments range from 11 % at Isara to 39 % at Malka. Selected field characteristics are summarized in Table 3.5.

For the selected sites, farmers reported an olive tree age of around 25–35 years. As they replace dead trees with younger seedlings, only orchards with trees of the same age were used.

All sites were conventionally tilled by animal power 2–4 times a year. This is done at least (i) immediately after the summer harvest in October/November, (ii) before the rainy season, and (iii) it might be repeated in February and March. Furrows follow up and down the hill and are interrupted by 1–4 lines of contour ploughing between each olive tree row down the hill. At all sites there are signs of siltation, elution of rocks, and rills of erosion visible in the furrows.

Table 3.5 Characteristics of the seven fields

Study area	Area (m^2)	Recent slope (°)	Aspect	Trees (no.)	Tree density (no. trees ha^{-1})	Rock fragments (%)	Texture[a]
Sifin II	535	16.1	SW	15	280	23	Silty clay loam
Tall Zira'a	771	5.8	WSW	16	208	20	Silty clay loam
Rusan	1213	6.7	N	19	157	17	Silt loam
Isara	709	5.7	W	24	339	11	Silt loam
Sifin I	651	11.8	SW	15	230	26	Silty clay loam
Sum	404	6.0	ENE	15	371	23	Silty clay loam
Malka	478	16.2	SW	14	293	39	Silty clay loam

[a]According to FAO (1979)

3.2.3 *Methods*

3.2.3.1 Principle of Method—Mound Development

Soil mounds under trees are attributed to different formation processes that can be separated into accumulative or erosive processes, as listed by Bochet et al. (2000) and Dougill and Thomas (2002). Accumulative processes are encompassed by the sedimentation of aeolian sediment under shrubs or splash erosion leading to more particles being moved towards the area under the canopy. Furthermore, human-induced accumulative forms are known from the research area where farmers create water-holding dams around stems to collect the overland flow near the tree or the accumulated organic fertilizer directly around the stems (Fig. 3.7e). As an erosive process, Bochet et al. (2000) mention water erosion that leaves the mounds as remnants of the former soil surface. Vanwalleghem et al. (2010) add tillage as an anthropogenic erosive process (Fig. 3.7a–c).

To use mounds as an indicator of surface lowering, it must be made certain that the mounds under investigation are residual erosion markers and not anthropogenically accumulated, as done by some farmers in Jordan for fertilization but not when planting the seedlings. However, in the field, residual erosion mounds could be clearly distinguished from other human-induced forms and the choice was validated through interviews with the farmers. Selection criteria included:

- Visible signs of the uncovering of roots due to erosion. Broken roots protrude from the mounds.
- The mounds show a naturally developed granular to sub-angular blocky structure, whereas the soil between the mounds is much less structured and tillage has destroyed the soil aggregates.

Fig. 3.7 Olive mounds used for measurement (**a–c**), and accumulative forms not used (**d–f**)

- The height of the mounds depends on the local steepness and reaches the maximum above the concave working edge, corresponding to the location where the highest transport energies are reached by the runoff, and eventually erosion rates are the highest (Van Oost et al. 2005).

The growing of the mound through edaphon activity or a dense root matrix, which decreases the bulk density of the surrounding soil and lifts the mound as described by Biot (1990) and Bochet et al. (2000), were additionally considered. Nevertheless, mounds were never observed in areas with declinations lower than 4° and therefore this possibility was excluded.

The orchards selected for investigation showed at least an inclination of 5–6°. This is required for a reliable differentiation of the mounds that are formed as a result of the earth banks caused by the plough. Furthermore, no small-scale terracing and the common ploughing practice are a prerequisite for the formation of the mound and thus for the application of the method. Hence, the obtained results are representative only of such positions in the relief and reflect areas of high erosion rates. However, these characteristics are quite representative of the catchment, and erosion features also exist on fields with less slope.

3.2.3.2 Measurements of Mounds and Soils

The topographic measurement of the olive orchards was done with a level tube TopCon G64, offering a theoretical vertical accuracy of ±2 mm km^{-1}, which is the most sensitive value in this research. Furthermore, it is not dependent on electrical power or satellite connection, both of which could not be secured during the work on-site (Ramos et al. 2008). However, the accuracy can be affected by solar calefaction of the tripod and therefore the distortion of the stand, refraction due to heating of the ground below the instrument, manual data transfer, the positioning of the measuring pole, and the inhomogeneous micro relief caused by the plough. The

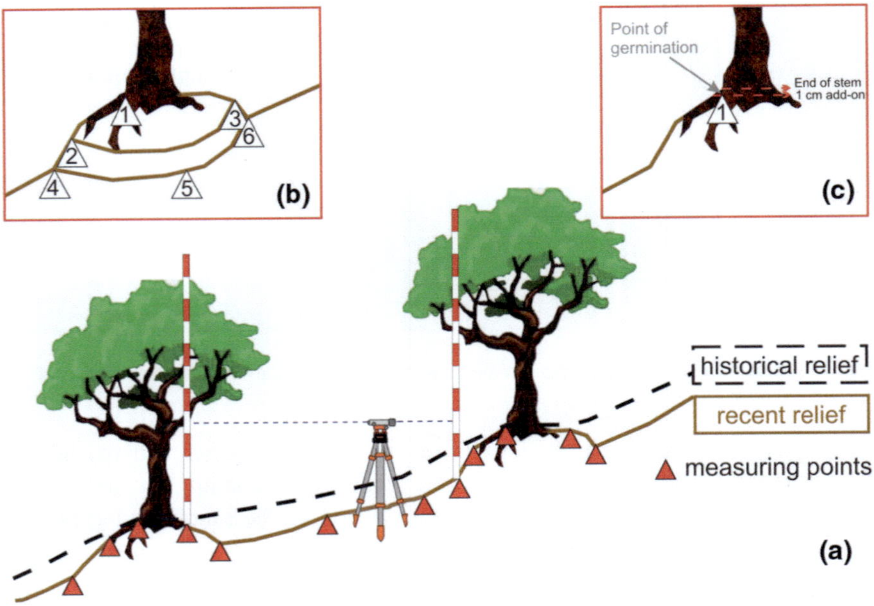

Fig. 3.8 Abstraction of the historical and recent relief (**a**) (level instrument by Fatangu at de. wikipedia [CC-BY-SA-2.0-de (http://creativecommons.org/licenses/by-sa/2.0/de/deed.en)], from Wikimedia Commons; modified). Enlargements: measurement points around the mound (**b**) and point of germination (**c**)

error was calculated with re-measurements of the instrument position with an average of 1.24 cm.

On each site, ca. 150–200 points were measured depending on the form of the olive mound. In the example of the Rusan field it is observable that most mounds are solitary developed (Figs. 3.7b and 3.9b). In deviance to the conic mound shapes measured by Vanwalleghem et al. (2010), the mounds in Jordan existed more as a polygonal plateau rather than a conic form. Therefore, a different measurement pattern was used to guarantee an improved representation of the mounds as a good base for the erosion calculation. Their shape was measured by taking points (Fig. 3.8b): on top (point 1), on half slope (points 2 and 3), and in the direction of the runoff before and after the mound (points 4 and 6) as well as on each side (points 5 and 7—opposite point 5, not visible in Fig. 3.8b). The mound shape measurements were supplemented by additional points in the middle between adjacent mounds and on the edges of the fields.

If the erosion could be traced by the uncovering of the roots in the direction of the slope as seen in Fig. 3.7c, fewer measurements were taken upslope. Measurements on top of the mound, where the olive stems appear, represent the point of germination of the tree and hence the historical relief when the orchard was planted. However, during the field measurements it was observed that olive roots in the ground do not have bark, but once exposed to the air they develop bark of about

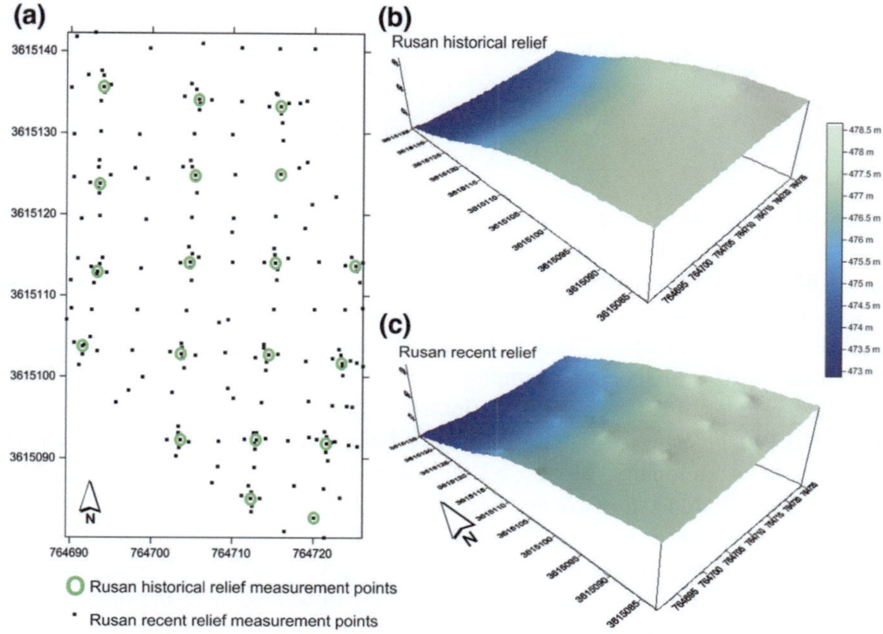

Fig. 3.9 Sampling scheme of Rusan site (projection: UTM, WGS84, zone 36 N). **a** Sampled field with measurement points around the mounds covering all four directions following Biot (1990). **b** Interpolated points of germination for the historical relief generation. **c** Interpolated recent relief measurement points

1 cm thick to protect against transpiration. This was evaluated by coring several exposed roots and by measuring the bark. Knowing this, measurement point 1 (Fig. 3.8b, c) was taken 1 cm below the determined point of germination (Fig. 3.8c) to take account of the subsequent thickening of the bark, consequently resulting in a slightly higher point of germination.

Attention has to be paid to olive trees that grew adventitious roots. These shoots grow along the stems during an intermediate accumulation of sediment and can be misinterpreted as the original roots, leading to a substantial overestimation of erosion. However, they are easy to distinguish. Orchards showing these or other anthropogenic-induced forms were excluded.

Afterwards, relative TopCon measurement points were transformed into real UTM coordinates (UTM, WGS84, zone 36 N). Raster datasets for the recent and the historical relief using only the top measurement points were respectively generated with the Natural Neighbourhood method in SURFER 9 (Golden Software). This method delivered the smallest root mean square errors (RMSE) for all fields in comparison to other interpolation methods such as Kriging, TIN, IDW, and Nearest Neighbour and is appropriate for spatial data sets that show heterogeneous distributed data records, as was the case here. The eroded volume was then calculated via the “Cut and Fill” option in Surfer 9 on a raster basis and transferred to an average soil loss rate *SL* (t m^{-2} $year^{-1}$) (Eq. 3.1).

$$SL = \frac{\rho_b}{t\Omega} \sum_{x_i}^{n} a_x [z_h(x) - z_r(x)] \quad (3.1)$$

where *SL* = soil loss averaged over the time of the orchard existence (t m^{-2} $year^{-1}$); ρ_b (t m^{-3}) = bulk density of the intra-mound sample; *t* = age of the orchard (year); Ω = surface of the orchard (m^2); a_x = surface elements within Ω (m^2); z_h(x); z_r(x) = historical and recent elevations at location *x*, respectively (m).

The interpolated surfaces were analyzed at a resolution of 0.09–0.28 m^2. ρ_b was measured for 81 soil specimen samples from the upper 30 cm of the soil surface. Samples were taken between the trees (intra-mound) and from the mound. Between 10 and 13 samples were tested from each field, depending on size and shape (Table 3.6). Additionally, stones >2 mm were removed through sieving. The average density of the stone content was determined using their weight and volume, which was measured using the water level method. The stone bulk density was then subtracted from the actual bulk density of the fine material following Lal and Shukla (2004). For the *SL* calculation, the average bulk density of the intra-mound samples was used for each field because it is assumed that the upper soil layer is commensurate with the already eroded soil.

On each field two olive trees were cored (*n* = 14), additionally one root strand and two wooden disks could be tested. The number and width of tree rings were measured using a stereo microscope (STEMI 2000C) and repeated with the Image Analyzer and entered into TSAPwin Scientific Version 0.55 by RINNTECH to display and compare the tree growth over the years. Since winter temperatures in Wadi Al-Arab usually fall below 5 °C, there is only one vegetation period, from March until November, resulting in one annual growth ring per year (Smith and Lewis 2007; Lucke 2007). Nevertheless, trees can develop “pseudo growth rings” that cannot be related to their age. The coring results of each orchard corresponded ±3 years to estimated ages reported by the farmers. For the erosion calculation the measurements were used.

Soil texture, field capacity, and saturated hydraulic conductivity (K_{sat}) values were measured to derive additional information on existing differences between mound and intra-mound positions. The soil texture was established from 37 samples (each field around 4–6 samples) by laser diffractometry with the Mastersizer S long bed Version 2.14 (Malvern Instruments Ltd). The field capacity (pF 1.8–63.1 hPa) and values were measured on the soil specimen of the three fields (Tall Zira’a, Rusan and Isara; *n* = 37) with the Sandbox (edition 08.01) and the Permeameter (edition 09.02) by Eijkelkamp (Giesbeek, Netherland), respectively. Results were tested in SPSS 19 and OriginPro 9.0G applying the independent *t*-test to check significant dependencies between the locations (mound or intra-mound) and the respective parameters (Webster 2001).

Table 3.6 Soil properties for mound and intra-mound positions with statistical indications

Study area	No. soil specimen, mound	No. soil specimen, intra mound	Avrg. bulk density, mound[b] (g cm^{-3})	Avrg. bulk density, intra mound[b] (g cm^{-3})	Avrg. K_{sat}, mound (m/d)	Avrg K_{sat}, intra mound (m/d)	Avrg. field capacity, mound (%)	Avrg. field capacity, intra mound (%)	Avrg. sand content, mound (%)	Avrg. sand content, intra mound (%)
Sifin II	5	5	1.0	1.1	n.a.	n.a.	n.a.	n.a.	19.7	5.8
Tall Zira'a[a]	0	13	n.a.	1.2	n.a.	0.7	n.a.	39.0	n.a.	9.3
Rusan	6	6	1.1	1.2	0.9	0.3	43.7	43.0	12.6	11.4
Isara	4	8	1.1	1.2	0.4	0.3	48.5	53.2	7.6	4.6
Sifin I	5	5	1.0	1.0	n.a.	n.a.	n.a.	n.a.	11.7	4.6
Sum	6	6	1.2	1.1	n.a.	n.a.	n.a.	n.a.	n.a.	n.a.
Malka	6	6	0.9	0.9	n.a.	n.a.	n.a.	n.a.	12.8	5.6
MEAN[c]			1.0 ± 0.1	1.1 ± 0.1	0.2 ± 0.1	0.1 ± 0.07	45.6 ± 4.6	43.3 ± 2.4	12.8 ± 6.	6.4 ± 3.3

[a]First sampling period/no sampies were taken from mounds
[b]Without stone content
[c]Averaged and independent *t*-test implemented without Tall Zira's samples

3.2.4 Results

The average bulk density from all mound samples (1.0 ± 0.1 g cm^{-3}) (not reached by the plough) and intra-mound samples (1.1 ± 0.1 g cm^{-3}) show that the intra-mound positions demonstrate a slightly higher density. This is also reflected through slightly higher average K_{sat} and field capacity values of the mound samples (0.2 ± 0.1 m d^{-1}; 45.6 ± 4.6 %) than those of the intra-mound positions (0.1 ± 0.1 m d^{-1}; 43.2 ± 2.4 %). However, the differences are not significant by means of the independent *t*-test. Only the average sand proportion of the mound samples is significantly higher (12.8 ± 6.4 %) than the means of intra-mound samples (6.6 ± 3.3 %; t (14.6) = 2.864, p = 0,012; Table 3.6) resulting in a decrease in the amount of clay and silt in the mound samples. Furthermore, if one considers each site individually, the fields Sifin II (t (10) = 2.429, p = 0.035) and Rusan (t (10) = 2.671, p = 0.023) have a significant difference in average bulk density. For the mean K_{sat} value this applies only for the Rusan site (0.27 ± 0.14 m d^{-1} mound; 0.09 ± 0.05 m d^{-1} intra-mound (t (9) = 3.079, p = 0.013).

The K_{sat} values for all fields are "relatively low" as classified by Landon (1991), supporting water ponding and surface runoff when it rains.

The calculated soil loss from each site ranges from 72 ± 7 to 135 ± 19 t ha^{-1} $year^{-1}$ with an average of 95 ± 8 t ha^{-1} $year^{-1}$ (Table 3.7). Results qualitatively indicate a relationship between orchard age and erosion rate. The age of the younger fields corresponds to the governmentally encouraged olive orchard expansion in the 1990s when they were expanded to steeper slopes and less suitable land. Although the sampling size does not support the inference statistically (r^2 = 0.4), it is believed that the relationship is plausible.

When plotting soil erosion depth versus slope for each pixel (Fig. 3.10) some pixels understandably show higher slope gradients (≥35°) than the average slope evaluated for the whole field, as stated in Table 3.5. An increasing mean erosion depth with slope can be observed in those pixels from 0° to 21° declination (−0.14 to −0.35 m).

Values remain around the same range between 21° and 35° (−0.31 to −0.34 m) and exceed the mean of −0.41 m erosion on a few very steep slopes (≥35°). Erosion plotted against profile curvature (Fig. 3.11) shows that areas now concave suffered from the strongest mean soil erosion. Concave positions also show the highest variability in the data, and most areas show a straight slope.

Figure 3.12 displays the computed erosion maps and the spatial distribution of soil loss across the slope for three out of the seven studied sites. It is visible that the different erosion patterns in the maps correspond well to the observed geomorphological erosion structures in the field. This demonstrates that both the measurements and the following interpolation are in line with reality, thus supporting the reliability of the applied method.

For natural erosion it is generally stated that the highest erosion impact takes place above the concave working edge, where runoff reaches the highest transport

Table 3.7 Parameters for erosion calculation listed by the age of the olive orchards

Study area	Area (m^2)	Area (ha)	Eroded volume (m^3)	Avrg. bulk density intra mound (g cm^{-3})	Standard deviation (g cm^{-3})	Tree age (year)	Yearly soil loss (t ha^{-1} $year^{-1}$)	±Error (t ha^{-1} $year^{-1}$)
Sifin II	535	0.05	136	1.1	0.16	21	**135**	**19**
Tall Zira'a	771	0.08	153	1.2	0.07	22	**107**	**6**
Rusan	1213	0.12	371	1.2	0.05	27	**132**	**6**
Isara	709	0.07	115	1.2	0.12	27	**72**	**7**
Sifin I	651	0.07	194	1.0	0.09	34	**87**	**8**
Sum	404	0.04	107	1.1	0.07	36	**83**	**5**
Malka	478	0.05	221	0.9	0.10	54	**77**	**9**
Mean	**680**	**0.07**	**185**	**1.1**	**0.09**	**32**	**95**	**8**

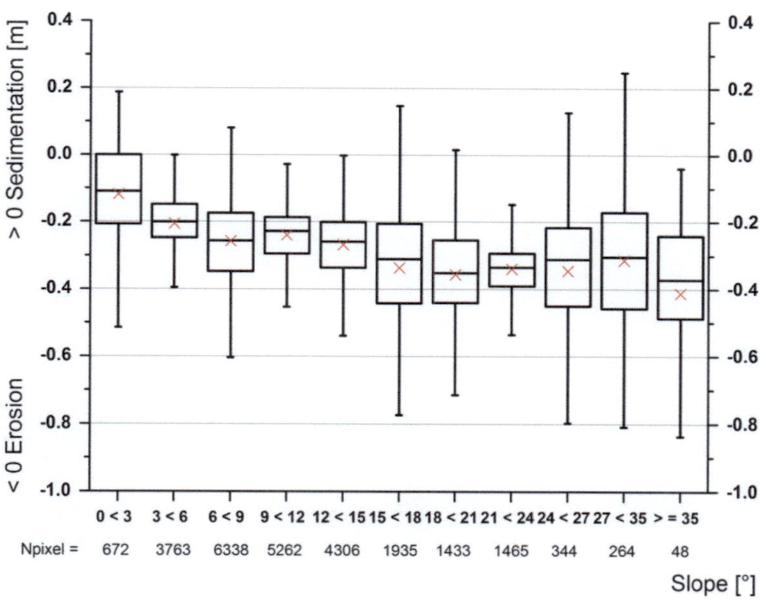

Fig. 3.10 Erosion depth plotted against the slope of the recent relief (pixel values of all fields, X = arithmetic mean)

energy (Van Oost et al. 2005), as is the case at the Sifin II site (Fig. 3.12). In contrast, at the Rusan site the slope increases immediately from 3 to 14 % in the middle of the field, augmenting the level of soil erosion. This is visible as a horizontal zone on the plotted slope shown in Fig. 3.12. According to the farmers, the reason for the slope increase was an intensive precipitation event that washed

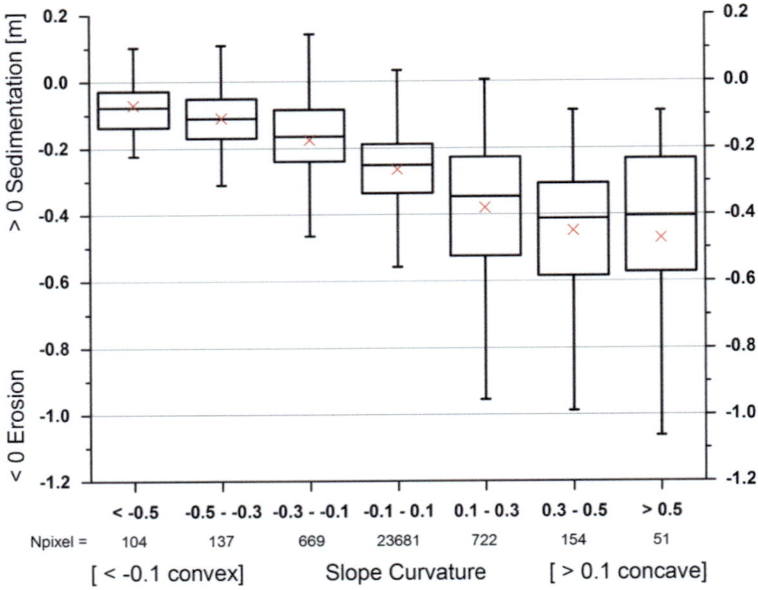

Fig. 3.11 Erosion depth plotted against the profile curvature of the recent relief (pixel values of all fields, X = arithmetic mean)

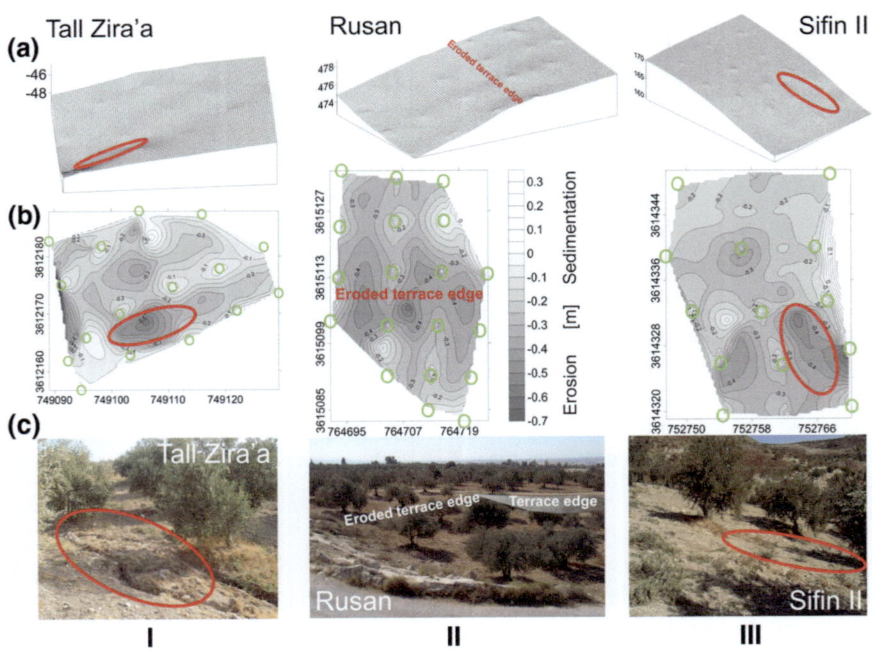

Fig. 3.12 Exemplary fields (projection: UTM,WGS84, zone 36 N) with **a** 3D surface forms, **b** soil erosion map (m), and **c** corresponding erosion features for sites I to III

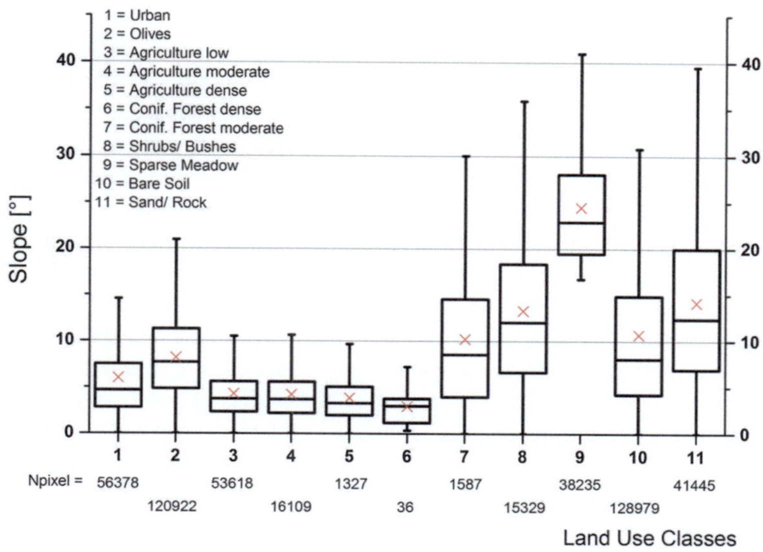

Fig. 3.13 Land use classes and their occurrence on slope positions (pixel values of the wholeWadi Al-Arab catchment, X = arithmetic mean)

away the terrace that secured the well-developed soil on the top. Tall Zira'a shows the patterns from the gully induced by water that accumulated on the street above the olive orchard and cut its way through the field (Fig. 3.12).

The investigation of the olive mounds was restricted to orchards with a steepness of at least 5–6°, which is required for a reliable differentiation of the mounds as described in the Methods. A simple GIS intersection of the land use map with the slope and soil information showed that ca. 19 % of the total catchment area shows these characteristics and therefore might also be of a similar high erosion potential as the studied fields. Additionally, if further expansion of olive fields occurs, they have to expand to yet unused fields and slopes. These potential areas of land suitable for olive plantations are currently captured in the land use classes Shrubs/Bush (8 in Fig. 3.13), Sparse Meadow (9), Bare Soil (10) and Sand/Rock (11) and show mean high slope gradients of between 10° and 24° as in Fig. 3.13.

3.2.5 *Discussion*

3.2.5.1 Results in the Scope of Soil Weathering and Literature

The erosion rates obtained are much higher compared to soil renewal rates by weathering and dust input as indicated by various authors. Verheijen et al. (2009) state that the rates of soil renewal are 0.3–1.4 t ha^{-1} $year^{-1}$ for Europe. Jürgens and

Fander (1993) refer to a rate of about 1 t ha^{-1} $year^{-1}$ from studies in Bavaria on shallow soils. For Middle Europe, Visser et al. (2004) give an estimate of tolerable soil loss of about 1.4 t ha^{-1} $year^{-1}$. Montgomery (2007) criticizes the high T value (soil-loss tolerance value) of 5–12 t ha^{-1} $year^{-1}$ established by the U.S. Department of Agriculture. He suggests a much lower mean value of 0.173 ± 0.029 mm $year^{-1}$ ($\sim$2.6 t ha^{-1} $year^{-1}$) of global geological erosion and therefore the natural equilibrium between soil production and erosion. Montgomery (2007) also argues that the soil production rate and geological erosion and the erosion value of conventionally tilled fields differ by at least one magnitude. Gómez et al. (2003) use a much higher threshold for Spain of between 8.4 and 11.4 t ha^{-1} $year^{-1}$, which varies according to soils (Moreira-Madueño 1991).

The results from NW Jordan exceed these soil loss rates at least 7 times, with an average of 95 ± 8 t ha^{-1} $year^{-1}$. Gómez et al. (2005) as well as Fleskens and Stroosnijder (2007) postulate that although 80 t ha^{-1} $year^{-1}$ has become an accepted average soil loss rate in olive orchards in Andalusia, this might be erroneous due to limited plot experiments and coarse scale estimates from the USLE model. Fleskens and Stroosnijder (2007) are of the opinion that, in general, average erosion rates are not likely to exceed 10 t ha^{-1} $year^{-1}$. Gómez et al. (2008b) criticize these conclusions and state that frequently rates higher than 10 t ha^{-1} $year^{-1}$ are measured. Other studies with similar landscape characteristics and tillage practice in the Mediterranean region corroborate the high erosion rates found in this study: Mabit et al. (2012) applied the ^{137}Cs method and indicated that maximum erosion rates of 19 t ha^{-1} $year^{-1}$ on an olive orchard with an average slope of 13 % is possible. Vanwalleghem et al. (2010) reported average erosion rates of 95 t ha^{-1} $year^{-1}$ using the mound method, and Vanwalleghem et al. (2011) obtained rates of 8–124 t ha^{-1} $year^{-1}$ based on the RUSLE model. In runoff plot studies, Bruggemann et al. (2005) measured soil losses of up to 81 t ha^{-1} $year^{-1}$. Even sloping orchards under vegetation succession may show erosion rates of 80 t ha^{-1} $year^{-1}$ (Stroosnijder et al. 2008).

The above observations lead to the conclusion that (1) until now there is no commonly accepted value for the potential erosion rates, and (2) our high erosion results lie within the scope of previous findings.

3.2.5.2 Processes of Erosion and Mound Formation

The mounds are often larger than the actual root zone and have a polygonal shape. Together with broken roots that protrude from the side walls, these characteristics indicate that the initial form is shaped by the farmers who omit them while ploughing up and down the hill. This is supported by the analysis of the mound and intra-mound soil samples—they differ significantly in grain size with mound samples having a higher sand proportion and intra-mound samples generally showing higher soil density, lower K_{sat} and field capacity values as well as no naturally developed soil architecture. The latter can be attributed to the destruction of soil structure through soil compaction as a result of ploughing and traffic (He et al. 2009).

The low K_{sat} values and the clayey soils indicate that surface runoff is likely to occur fast during intense and/or long-lasting precipitation events. Farmers reported strong damages during rainfall events especially in 1992 and 2003. Rills, gullies, and siltation along the furrows of the plough circumvent and define the mounds and uncover roots (Figs. 3.14 and 3.15), as reported by Vanwalleghem et al. (2010). To conclude, the olive mounds in Wadi Al-Arab are the consequence of tillage and runoff.

In addition, erosion rates were higher in younger fields cultivated on marginal lands, suggesting that smaller olive canopies leave the soil less protected (Gómez et al. 2009).

Nevertheless, according to Eltaif et al. (2010), the average rain erosivity factor for USLE and RUSLE for the Wadi Al-Arab is relatively low, indicating that processes not induced by rainfall might be important in NW Jordan (Eltaif et al. 2010). However, no geomorphological indicators for wind erosion were found, and clayey soils like those in the study area are resistant against it.

Fig. 3.14 Olive mound on field Sifin II

Fig. 3.15 Furrows canalizing the runoff and defining the mounds

Vanwalleghem et al. (2010) ascribe nearly 50 % of the soil loss in one site to tillage translocation. Van Oost et al. (2005) and García-Ruiz (2010) mention that the mechanization of agriculture plays a major role in soil erosion processes. In contrast, tracer experiments by Barnevald et al. (2009) show that even with simple donkey-drawn tillage equipment, which is typically used in the study area, soil loss of 80 t ha^{-1} $year^{-1}$ is possible on land with 25 % average slope and five up- and down-tillage passes per year. This alone can explain the erosion rates of most orchards considered in this work, which is consistent with the statement made by García-Ruiz (2010) that "soil displacement by ploughing should be at least one order of magnitude greater than that due to water erosion in similar environments".

3.2.5.3 Implications and Mitigation Strategies for Land Management

Several features in the study area including (i) steep valleys with only a few recognizable alluvial deposits, (ii) lack of thick bed material in wadis, and (iii) gullies that are cut into the bedrock point to a strong connectivity between the hillslopes and the Wadi Al-Arab reservoir. According to Chap. 4, the reservoir annually receives around 67,000 $m^3 \pm 6$ % of eroded sediments, threatening the life of the reservoir which feeds irrigated plains in the Jordan Valley and the city of Amman (Wadi Al-Arab dam director Hussein Al-Shurieki, personal communication).

Preventive measures against slope erosion would challenge the farmers' tradition of up and down tillage. In interviews, local farmers stated that tillage increases soil porosity and infiltration to assure a better delivery of nutrients to trees. Increased surface roughness due to ploughing may facilitate this (Henderson 1979; Gómez et al. 1999, 2004; Fleskens and Stroosnijder 2007). Furthermore, growing weeds may reduce water supply to olives trees, provide a space for poisonous animals, and offer food for grazing sheep and goats that also might feed from the olives. However, various studies including the present one also indicate that compared to conventional tillage, no tillage has positive effects on soil hydraulic properties. Azooz and A'Arshad (1996) highlighted the positive effects of no tillage on soil pore structure which significantly increased hydraulic conductivity for sites in Canada. He et al. (2009) found a comparable result especially in soil profiles of 20–30 cm deep with a significantly lower bulk density, higher infiltration, and water retention on non-tilled fields on the Loess Plateau in China. Gómez et al. (2008b) describe that increased infiltration by tilling is only temporary and eventually causes higher runoff than a cover crop.

Dabney et al. (2001), Hartwig and Ammon (2002), and Gómez et al. (2005, 2008a) indicate that cover crops compete with trees for water in semi-arid regions. Nevertheless, Hartwig and Ammon (2002) and Barnevald et al. (2009) also point out the advantages of growing a cover crop such as nutrient enrichment, pest control, and less soil erosion. Hernández et al. (2005) found that an understory plant cover system does not significantly reduce soil water content compared to conventional tillage and showed a beneficial effect of clover enriching the surface soil layer with water, carbon, nitrogen, and nitrate. Hartwig and Ammon (2002) recommend cover crops in orchards for weed control, better nitrate management, and higher productivity.

Therefore, some simple changes in the management of these orchards such as contour tillage (Barnevald et al. 2009) could reduce erosion. Additional terracing on steep slopes, fewer ploughing passes per year, and semi-permanent intercrops may stabilize the soil layer, recycle nutrients, produce organic matter, and thus raise the field capacity, attracting beneficial insects, and saving working hours (Bruggeman et al. 2005; Gliessman 2007).

Olive farmers witnessed an increasing number of dying trees on steep slopes (≥11.5°) in the years of investigation and explain this phenomenon by the intensified pumping of ground water and by the falling water table in the region. However, the absence of nutrient rich topsoil, accelerated runoff, less infiltration and therefore less water availability, and tree diseases such as Verticillum may also be responsible. This calls for more investigations in relation to intensification and mechanization of agriculture (Jacquet et al. 2007), since fields operated by heavy machinery require trees to be planted further apart, thus leaving more soil unprotected. Such agricultural intensification caused soil erosion resulting from the response to the EU Common Agricultural Policy and the subsidies to olives throughout the 70s, 80s, and 90s (García-Ruiz 2010; Vanwalleghem et al. 2011).

In Jordan, water and soil resources have been, until now, monitored separately. The sectorial management by Jordanian authorities seems counterproductive in the attempt to tackle the challenges concerning the demand for fresh water in the region. Soil needs to be considered as a finite resource and in direct interrelation to water supply and quality.

3.2.6 Conclusion

The research based on seven olive orchards shows that the method proposed by Vanwalleghem et al. (2010) gives reliable results for fields with 5° to 6° slope in NW Jordan. The soil loss ranges between 72 ± 7 and 135 ± 19 t ha^{-1} $year^{-1}$ and is connected to the age of the orchard. Around 19 % of the catchment area likely has soil loss rates in this range. We conclude that the rapid erosion is due to the agricultural intensification since the 1990s and the expansion of fields on steeper marly slopes. This intensification occurred as a result of the governmental policies to provide jobs for people in the rural areas and to limit urbanization. With the 2014 European trading zone expansion to Jordan, the intensification will accelerate following a liberalized market. The sectorial management of resources may inhibit future effective land management.

3.3 Implementation and Results of 6 Erosion Pin Fields

Following the method described by Fanning (1994), Keay-Bright and Boardman (2009), and Higuchi et al. (2013), seasonal erosion rates on marly slopes were measured in 6 erosion pin fields of 200 m^2 (10 × 20 m). The pins were made from steel and were 15 cm long and had a diameter of 0.4 cm. At 10 cm, a 0.1 cm thick metal disc (1.5 cm in diameter) was welded to the pin (Figs. 3.1 and 3.16). Every 2 m a pin was carefully driven 10 cm into the earth until the lower side of the metal disc reached the surface, indicating the present surface level.

In total 66 pins per field were installed in March 2011 and October 2011. The fields covered different exposed hills and were situated halfway downslope. All fields showed natural shrub vegetation on silt loam with an average of 17 % clay, 71 % silt, and 13 % sand (FAO grain size classification 1979). Fields exposed to the south showed a more sparse vegetation cover and more erosive features than those exposed to the north (Sect. 2.3).

The erosion pins in fields 1–3 were installed in February 2011 (1–3) and read out in October 2011, covering the rainy spring of 2011. In the following autumn and winter season from November 2011 to May 2012, fields 1 and 2 were read out a second time and new pins were installed in fields 4–6 and read out as well (Fig. 3.16a–c). The seasons showed summarized rain values of 231 and 379 mm respectively (own data).

Measurements were done in the case of erosion from the bottom of the metal disc to the lowered surface using a metal measuring tape and to the nearest

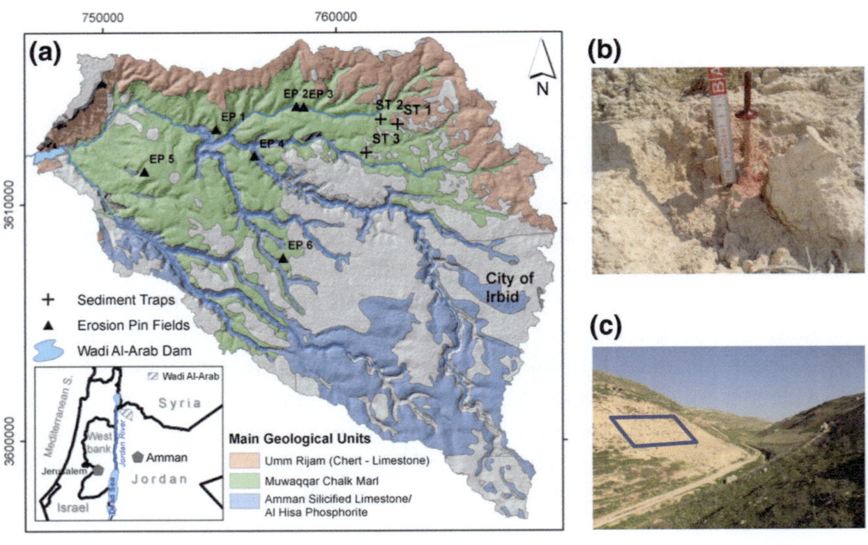

Fig. 3.16 Erosion pin fields in the Wadi Al-Arab catchment. **a** Location of erosion pin fields (EP 1–6) and sediment traps (ST1–3), **b** uncovered pin in EP 1, **c** EP 3, SSW exposed (MWI data, 2011; projection: UTM, WGS84, zone 36 N)

millimetre. If pedestals developed under the metal disc, measurements were implemented to the base of the pedestal. In the case of sedimentation, the pin was located with a Bosch DMF 10 Zoom wire tracer, carefully excavated, and the thickness of the material was measured from the bottom of the metal disc. To account for differences in measurements due to surface roughness, 1 mm measurement error was assumed for the calculation.

Some of the pins got knocked down by grazing sheep and were consequently neglected. If they were trodden on and pushed into the ground by the animals, the metal disc left a cylindrical imprint in the soil which was clearly detected during excavation. These pins were treated as drop outs. Except for EP 1, 30 pins were stolen, drop out ranged between 0 and 16 % (=0 to 10 pins).

The former and recent surfaces were then interpolated using SURFER 9 (Golden Software). The interpolation method "Natural Neighbour" was chosen following the smallest root mean square error in comparison to other methods such as TIN, IDW, Ordinary Kriging, Universal Kriging, and Nearest Neighbour. Finally, the two DEMs were subtracted from each other and the eroded/sedimented volume was calculated using soil density taken from 9 soil bulk density samples with an average of 1.26 ± 0.12 g cm^{-3} for the fields. Results, of course, are only representative for areas with similar characteristics, such as slope, soil type etc. (Table 3.8), and for the specific seasonal precipitation conditions (Keay-Bright and Boardman 2009).

3.3.1 Results and Discussion

For the time of observation, the mean amount of sediment eroded in the six fields results in 23.0 ± 5.4 t ha^{-1}. However, the sedimentation rates in the plots indicate that in the majority of cases more sediment was deposited than eroded in the pin fields. As the plots are not fenced on the top, sediments eroded above the fields can accumulate in the plots. When comparing the averages of south (30.6 ± 16.76 t ha^{-1}) to north (10.3 ± 12.1 t ha^{-1}) exposed fields, the means differ and confirm the observation described in Sect. 2.3. However, these differences are not statistically significant and a statistical argumentation in any case is problematic due to the small and unequal sample size.

Figure 3.17 shows clearly that highest erosion rates (6.5 cm) occurred in small concave structure, indicating rill erosion as the dominant erosion process. Highest sedimentation lay around 3.5 cm. On average, a mean disposition of 1 mm could be calculated for all the analysed slopes. Highest erosion rates were reached on slopes with 24° to <26°.

It was observed several times that organic pieces transported by overland flow or gravital forces got trapped at the pin top and then sediments accumulated upslope. Additionally, one pin (Fig. 3.16b) was observed to be situated at a rill head. The reason for this might be that the pins concentrate overland flow and thus cause rill development. However, further studies would have to investigate this phenomenon as it was only observed once out of 340 pins.

Table 3.8 Erosion pin field characteristics, erosion and sedimentation rates, and calculated eroded volume per hectare

Field	Rain season	Aspect	Slope (°)	Area (m^2)	Failure of pins (No./%)	Volume eroded (m^3)	Volume sed. (m^3)	Volume eroded (t ha-1)	Error (t ha^{-1})
EP 1	02.′11–10.′11	SW	32.4	80.0	30/46	0.34	0.08	53.5	5.1
EP 1	11.′11–05.′12	SW	32.4	80.0	30/46	0.27	0.24	42.5	4.0
EP 2	02.′11–10.′11	NNW	21.2	170.9	7/11	0.33	0.42	24.3	2.3
EP 3	02.′11–10.′11	SSW	32.4	200.0	0/0	0.39	1.00	24.6	2.3
EP 3	11.′11–05.′12	SSW	32.4	186.5	3/5	0.26	1.38	17.6	1.7
EP 4	11.′11–05.′12	NE	31.8	159.0	10/16	0.04	1.06	3.2	0.3
EP 5	11.′11–05.′12	NW	27.3	186.5	3/5	0.05	0.34	3.4	0.3
EP 6	11.′11–05.′12	S	39.2	188.1	3/5	0.22	0.97	14.7	1.4
							Mean	**23.0**	**±17.7**

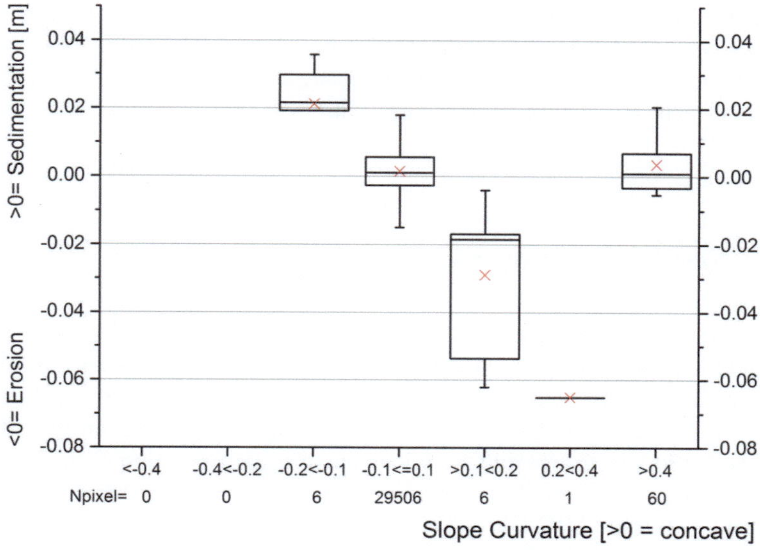

Fig. 3.17 Erosion and sedimentation results plotted against profile curvature of the recent relief of all the analysed erosion pin fields

3.4 Building of Event Based Sediment Traps and the Assessment of a Few Events

Slopes with consolidated limestone and soil patches are abundant in the catchment as described in Sect. 2.3. To analyse whether rain events have an erosive effect on the soil patches on hard limestone, 3 sediment traps were masoned on 3 different slopes in February 2011 after consultation with landowners and nearby Beduins. Therefore, soil patches with indications of a concentrated outflow were chosen on 3 different hills with slopes ranging from 5–12° (Fig. 3.18a–c).

Fig. 3.18 Sediment traps 1 (**a**), 2 (**b**), and 3 (**c**) after construction

The small catchments (soil patches) were isolated from their surrounding with a cemented wall, and the concentrated runoff from the patches was collected in a cemented bucket or a masoned basin with plastic sheeting. All the patches showed a silt loam with an average of 1 % clay, silt 80 %, and 9 % sand (classification FAO 1979).

3.4.1 Results and Discussion

After revisiting the constructed sediment traps in October 2011, all of them showed signs of tampering to the point of complete destruction in the case of sediment traps (ST) 2 and 3. The cemented walls were perished and the plastic sheeting completely cut out of the basin. ST 1 showed only limited signs of destruction by shot gun firing and could be used for a total of 3 events to quantify erosion of the soil patch in November 2011 (Table 3.9). The area of the patch was measured with a geo-referenced picture in ARC GIS 9.3.

The soil patches showed a soil thickness of 6 cm up to 80 cm, when developed in karstic concave forms. The colour (7.5 YR 2.5/2-4) of the soil patches indicates that the soil had time to develop on site and thus points towards almost geomorphological stable positions. Nevertheless, erosion values showed that in the case of stronger rain events, the soil patches eroded between 0.05 and 0.29 mm m^{-2} per event and 0.49 mm m^{-2} in total in November 2011.

However, due to the destruction of the sediment traps and thus the limited number of samples and recorded events, the results allow only the conclusion that even on these hard limestone slopes with soil patches runoff and erosion occurs, even though one might assume that the soil patches themselves indicate a geomorphological stable position. The average erosion value of 1.79 ± 1.3 t ha^{-1} $event^{-1}$ refers only to soil patches and is not calculated for the whole slope because the uncovered limestone areas are not considered in the calculations.

Results were not used in the further course of the thesis due to the mentioned circumstances.

Table 3.9 Calculated erosion values from ST 1 for three rain events in November 2011

Rain events 2011	Rain amount (mm)	Aspect	Slope (°)	Area of drained patch (m^2)	Volume eroded (g m^{-2})	Volume eroded (t ha-1)	Height eroded (mm)
6 Nov	33.3	SW	12.6	3.52	54.38	0.54	0.05
18 Nov	26.3	SW	12.6	3.52	317.29	3.17	0.29
21 Nov	11.7	SW	12.6	3.52	166.77	1.67	0.15
					Mean	**1.79 ± 1.32**	

References

Adamiec G, Aitken M (1998) Dose-rate conversion factors: update. Ancient TL 16:37–50

Ad-hoc-Arbeitsgruppe (AG) Boden (2005) Bodenkundliche Kartieranleitung. Bundesanstalt für Geowissenschaften und Rohstoffe in Zusammenarbeit mit den Staatlichen Geologischen Diensten, 5th ed. Hannover

Alsharhan AS, Nairn AEM (1997) Sedimentary basins and petroleum geology of the middle east (Volume 1). Elsevier Science, Amsterdam (e.V)

Azooz RH, A' Arshad H, (1996) Soil infiltration and hydraulic conductivity under long-term no-tillage and conventional tillage systems. Can J Soil Sci 76:143–152

Barnevald RJ, Bruggeman A, Sterk G, Turkelboom F (2009) Comparison of two methods for quantification of tillage erosion rates in olive orchards of north-west Syria. Soil Tillage Res 103:105–112

Biot Y (1990) The use of tree mounds as benchmark of previous land surfaces in a semi-arid tree savanna, Botswana. Veg Erosion 26:437–450

Bruggemann A, Masri Z, Turkelboom F (2005) Strategies to sustain productivity of olive groves on steep slopes in the northwest of the Syrian Arab Republic. In: Benites J, Pisante M, Stagnari F (eds) Integrated soil and water management for orchard development; role and importance. FAO land and water Bulletin 10, FAO, Rome, pp 75–87

Bührig C (2008) Gadara/Jadar/Umm Qais. In: Bartl K, Al-Razzaq Moaz K (eds) Residences, castles, settlements. Transformation processes from late antiquity to early Islam in Bilad ash-Sham. Marie Leidorf, Rahden, pp 97–115

Dabney SM, Delgado JA, Reeves DW (2001) Using winter cover crops to improve soil and water quality. Commun Soil Sci Plant Anal 32(7):1221–1250

De Santisteban LM, Casalí J, López JJ (2006) Assessing soil erosion rates in cultivated areas of Navarre (Spain). Earth Surf Proc Land 31:487–506

Döring M (2009) Der längste Tunnel der antiken Welt. Antike Welt 2:26–34

Döring M (2010) 100 km unter Tage- Längster bisher bekannter Aquädukttunnel der Antike in Jordanein und Syrien. Bergbau 5:211–216

Dougill AJ, Thomas AD (2002) Nebkha dunes in the Molopo Basin, South Africa and Botswana: formation controls and their validity as indicators of soil degradation. J Arid Environ 50:413–428

Eltaif NI, Gharaibeh MA, Al-Zaitawi F, Alhamad MN (2010) Approximation of rainfall erosivity factors in North Jordan. Pedosphere 20(6):711–717

Euro-Mediterranean Agreement 2002/357/EC (2002) EU Association Council proposal on the implementation of Article 95 and article 84 of the euro-mediterranean agreement between EU member states and the kingdom of Jordan: http://eur-lex.europa.eu/Notice.do?mode=dbl&lang=de&ihmlang=de&lng1=de,fr&lng2=da,de,el,en,es,fi,fr,it,nl,pt,sv,&val=278622:cs&page=, Last checked 01 Nov 2013

Fanning P (1994) Long-term contemporary erosion rates in an arid rangelands environment in western New South Wales, Australia. J Arid Environ 28:173–187

Food and Agriculture Organization of the United Nations (FAO) (2006) Guideline for soil description, 4th edn. Publishing Management Service, Rome

Fleskens L, Stroosnijder L (2007) Is soil erosion in olive groves as bad as often claimed? Geoderma 141:260–271

Food and Agriculture Organization of the UN (FAO) (1979) Soil classification http://www.fao.org/soils-portal/soil-survey/soil-classification/en/, Last checked 12 June 2014

Francia Martínez JR, Durán Zuazo VH, Martínez Raya A (2006) Environmental impact from mountainous olive orchards under different soil-management systems (SE Spain). Sci Total Environ 358:46–60

Fuchs M, Straub J, Zöller L (2005) Residual luminescence signals of recent river flood sediments: a comparison between quartz and feldspar of fine- and coarse-grain sediments. Ancient TL 23:25–30

García-Ruiz JM (2010) The effects of land uses on soil erosion in Spain: a review. Catena 81:1–11

García-Ruiz JM, Nadal-Romero E, Lana-Renault N, Beguería S (2013) Erosion in mediterranean landscapes: changes and future challenges. Geomorphology 198:20–36

Gibson S (2001) Agricultural terraces and settlements expansion in the highlands of early iron age palestine: is there any correlation between the two? In: Mazar A (ed) Studies in the archaeology of the Iron Age in Israel and Jordan. Sheffield Academic Press Ltd, Sheffield, pp 113–148

Gliessman SR (2007) Agroecological principles for making the conversion to organic olive agroecosystems. In: Vossen PM (ed) 2007. Organic olive production manual. University of California. Agriculture and Natural Resources Publication. Global Water Partnership, Technical Advisory Committee (TAC) 2000. Integrated Water Resource Management, TAC Background Papers No. 4. Global Water Partnership, Sweden, pp 8–50

Glück N (1942) Further explorations in Eastern Palestine. BASOR 86:14–24

Gómez JA, Giráldez JV, Pastor M, Fereres E (1999) Effects of tillage method on soil physical properties, infiltration and yield in an olive orchard. Soil Till Res 52:167–175

Gómez JA, Battany B, Renschler CS, Fereres E (2003) Evaluating the impact of soil management on soil loss in olive orchards. Soil Use Manag 19:127–134

Gómez JA, Romero P, Giráldez JV, Fereres E (2004) Experimental assessment of runoff and soil erosion in an olive grove on a Vertic soil in southern Spain as affected by soil management. Soil Use Manage 20:426–431

Gómez JA, Romero P, Giráldez JV, Fereres E (2005) Water erosion in olive orchards in Andalusia (Southern Spain): a review. Geophysical Research Abstracts 7, European Geoscience Union 2005. SRef-ID; 1607–7962/gra/EGU05-A-08406

Gómez JA, Amato M, Celano G, Koubouris GC (2008a) Organic olive orchards on sloping land: more than a specialty niche production system? J Environ Manage 89:99–109

Gómez JA, Giráldez JV, Vanwalleghem T (2008b) Comments on "Is soil erosion in olive groves as bad as often claimed?" by J Fleskens and L Stroosnijder. Geoderma 93:93–95

Gómez JA, Sobrinho T, Giráldez JV, Fereres E (2009) Soil management effects on runoff, erosion and soil properties in an olive grove of Southern Spain. Soil Till Res 102:5–13

Grove AO, Rackham O (2001) The nature of mediterranean Europe, an ecological history. Yale University Press, New Haven

Hartwig NL, Ammon HU (2002) Cover crops and living mulches. Weed Sci 50:688–699

He N, Wang Q, Li H, Tullberg JN, McHugh AD, Bai Y, Zhang X, Mc Laughlin N, Gao H (2009) Soil physical properties and infiltration after long-term no-tillage and ploughing on the Chinese Loess Plateau. N Z J Crop Hortic Sci 37:157–166

Henderson DW (1979) Soil management in semi arid environments. Agric Semi-Arid Environ Ecol Stud 34:224–237

Hernández AJ, Lacasta C, Pastor J (2005) Effects of different management practices on soil conservation and soil water in a rainfed olive orchard. Agric Water Manage 77:232–248

Higuchi K, Chigira M, Lee DH (2013) High rates of erosion and rapid weathering in a Plio-Pleistocene mudstone badland, Taiwan. Catena 106:68–82

IOC (International Oil Council) (2005) Country Profiles—Jordan E.108/Doc. no. 4, Update no. 31, Ref.: 18, 1–17. http://www.internationaloliveoil.org/, Last checked 29 Aug 2011

Jacquet F, Petit M, Tyner W, El Hadad F (2007) Deliverable D27 impacts of agricultural trade liberalization between the EU and mediterranean countries. Synthesis Report. EU-MED AGPOL Project No. SSPE-CT-2004–502457, Report, 1–42 (http://eumedagpol.iamm.fr/html/publications/prj_report/default.html, Last checked 11 July 2012)

Jürgens C, Fander M (1993) Soil erosion assessment by means of LANDSAT-TM and ancillary digital data in relation to water quality. Soil Tech 6:215–223

Keay-Bright J, Boardman J (2009) Evidence from field-based studies of rates of soil erosion on degraded land in the central Karoo, South Africa. Geomorphology 103:455–465

Keilholz P (2007) Die Zisternen der antiken Stadt Gadara (Umm Quais, Jordanien). In: Ohlig C (ed) Antike Zisternen. Schriften der Deutschen Wasserhistorischen Gesellschaft (DWhG) e.V., 9. Books on Demand GmbH, Norderstedt, pp 195–228

Keilholz P (2012) The ancient cisterns of the Hellenistic Gadara/Umm Qais (Jordan). Tübinger Archäologische Forschungen [unpublished]

Khresat SA, Rawajfih Z, Mohammad M (1998) Land degradation in north-western Jordan: causes and processes. J Arid Environ 39:623–629

Klein M (2007) Back to the water cisterns. In: Ohlig C (ed) Antike Zisternen. Schriften der Deutschen Wasserhistorischen Gesellschaft (DWhG) e.V., 9. Books on Demand GmbH, Norderstedt, pp 195–228

Kosmas C, Danalatos N, Cammeraat LH, Chabart M, Diamantopoulos J, Farand R, Gutierrez L, Jacob A, Marques H, Martinez-Fernandez J, Mizara A, Moustakas N, Nicolau JM, Oliveros C, Pinna G, Puddu R, Puigdefabregas J, Roxo M, Simao A, Stamou G, Tomasi N, Usai D, Vacca A (1997) The effect of land use on runoff and soil erosion rates under mediterranean conditions. Catena 29:45–49

Koulouri M, Giourga C (2007) Land abandonment and slope gradient as key factors of soil erosion in mediterranean terraced lands. Catena 69:274–281

Kraushaar S, Herrmann N, Ollesch G, Vogel H-J, Siebert C (2014) Mound measurements—quantifying medium term soil erosion under olive trees in Northern Jordan. Geomorphology 213:1–12

Kraushaar S, Ollesch G, Siebert C, Vogel H-J, Fuchs M (2015) Long-termsediment export estimates from Northern Jordan using Roman cisterns as sedimenttraps. Geoarchaeology 30:369–378

Landon JR (ed) (1991) Booker tropical soil manual: a handbook for soil survey and agricultural land evaluation in the tropics and subtropics. Longman Scientific and Technical, Harlow

Lucke B (2007) Demise of the decapolis. Past and present desertification in the context of soil development, land use and climate. Dissertation at the Brandenburg Technical University of Cottbus

Mabit L, Chhem-Kieth S, Toloza A, Vanwalleghem T, Bernard C, Infante Amate JI, Gonzáles de Molina M, Gómez JA (2012) Radioisotopic and physicochemical background indicators to assess soil degradation affecting olive orchards in southern Spain. Agric Ecosyst Eviron 159:70–80

Mallast U (2011) Klassifikationsbericht des Smart Catchments (See Genezareth—unteres Jordantal —Totes Meer) basierend auf GLS 2010 Landsat ETM + Szenen. [unpublished]

Mauz B, Packman SC, Lang A (2006) The alpha effectiveness in silt-sized quartz: new data obtained by single and multiple aliquot protocols. Ancient TL, 24, 47–52

Mays LW (2013) Use of cisterns during antiquity in the Mediterranean region for water resources sustainability. Water Sci Technol Water Supply 14(1):38–47

Milgroom J, Auxiliadora Soriano M, Garrido JM, Gómez JA, Fereres E (2005) The influence of a shift from conventional to organic olive farming on soil management and erosion risk in southern Spain. Renewable Agric Food Syst 22(1):1–10

Ministry of Agriculture (MoA) (1994) National Soil Map and Land Use Project—The Soils of Jordan. Level 2. Volume 3. Hunting Technical Services LTD and Soil Survey and Land Reseach Center. Amman, Jordan

Ministry of Water and Irrigation (MWI), Jordan Valley Authority (JVA) (2011) Wild plants in jordan valley. Water Information and Public Awareness Department, Jordan

Mittmann S (1970) Beiträge zur Siedlungs- und Territorialgeschichte des nördlichen Ostjordanlandes. Otto Harrassowitz, Wiesbaden

Moh'd BK (2000) The geology of Irbid and Ash Shuna Ash Shamaliyya (WAQQAS). Map Sheets No. 3154-II and 3154-III, Bulletin 46. Geological Directorate, Geological Mapping Divisio, Amman

Montgomery DR (2007) Soil erosion and agricultural sustainability. Proc Nat Acad Sci United States of America 104(33):13268–13272

Moreira-Madueño J (1991) Capacidad de uso y erosion de suelos: una aproximación a la evaluación de tierras Andalucía. Agencia de Mesio Ambiente, Sevilla

Munsell Color (Firm) (1994) Munsell soil color charts. 1994 Rev. edn. Macbeth Division of Kollmorgan Instruments Corp., New Windsor (Munsell Color 1994)

Murray AS, Wintle AG (2000) Luminescence dating of quartz using an improved single-aliquot regenerative-dose protocol. Radiat Meas 32:57–73

Pastor M, Castro J (1995) Soil management systems and erosion. Sci Tech 59:64–74
Pérez FL (2007) Biogeomorphological influence of the Hawaiian silversword (Argyroxiphium sandwicense DC.) on soil erosion in Haleakala (Maui, Hawai'i). Catena 71:41–55
Porath J (1984) Lime plaster in Aqueducts. A new chronological indicator. Mitteilungen des Leichtweiss-Institut für Wasserbau 82:1–16
Prescott JR, Hutton JT (1994) Cosmic ray contributions to dose rates for luminescence and ESR dating: large depths and long-term time variations. Radiat Meas 23:497–500
Preusser F, Degering D, Fuchs M, Hilgers A, Kadereit A, Klasen N, Krbetschek M, Richter D, Spence JQG (2008) Luminescence dating: basics, methods and applications. Eiszeitalter und Gegenwart—Quarternary Sci J 57(1–2):95–149
Ramos MI, Feito FR, Gil AJ, Cubillas JJ (2008) A study of spatial variability of soil loss with high resolution DEMs: a case study of a sloping olive grove in southern Spain. Geoderma 148:1–12
Smith D, Lewis D (2007) Dendrogeomorphology. In: Elias SA (ed) Encyclopedia of quaternary science, 2nd edn. Elsevier, Amsterdam, pp 459–464
Stroosnijder L, Mansinho MI, Palese AM (2008) OLIVERO: the project analysing the future of olive production systems on sloping land in the mediterranean basin. J Environ Manage 89:75–85
Thornes JB (2005) Coupling erosion, vegetation and grazing. Land Degrad Dev 16:127–138
Van Oost K, Van Muysen W, Govers G, Deckers J, Quine TA (2005) From water to tillage erosion dominated landform evolution. Geomorphology 72:193–203
Vandaele K, Poesen J, Marques da Silva JR, Desmet PPJ (1996) Rates and predictability of ephemeral gully erosion in two contrasting environments. Géomorphol Relief Processus Environ 2(2):83–95
Vanwalleghem T, Laguna A, Giráldez JV, Jiménez-Hornero FJ (2010) Applying a simple methodology to assess historical soil erosion in olive orchards. Geomorphology 114:294–302
Vanwalleghem T, Infante Amte J, González de Molina M, Soto Fernández D, Gómez JA (2011) Quantifying the effect of historical soil management on soil erosion rates in mediterranean olive orchards. Agric Ecosyst Environ 142:341–351
Venot JP (2003) Volume IV: the peculiar case of the olive trees orchard. Reclamation's history of the Jordan River Basin in Jordan, a focus on agriculture: past trends, actual farming systems and future prospective. INA P-G Paris-Grignon National Institute of Agronomy. French Regional Mission for Water and Agriculture (MREA) French Embassy, International Water Management Institute (IWMI)
Verheijen FGA, Jones RJA, Rickson RJ, Smith CJ (2009) Tolerable versus actual soil erosion rates in Europe. Earth-Sci Rev 94:23–38
Vieweger D, Häser J (2010) Neue Ausgrabungen geben Auskunft über das Handwerk im biblischen Altertum. Welt und Umwelt der Bibel 3:68–69
Visser SM, Sterk G, Ribolzi O (2004) Techniques for simultaneous quantification of wind and water erosion in semi-arid regions. J Arid Environ 59:699–717
Webster R (2001) Statistics to support soil research and their presentation. Eur J Soil Sci 52:331–340
Yassoglou NJ, Kosmas C (2000) Desertification in the mediterranean Europe. A case in Greece. In: RALA (Ed) Rangeland Desertification Report, N 200. RALA Ed, Reykjavik, pp 27–33

Chapter 4
Qualitative Effects of Sediments on the Wadi Al-Arab Reservoir

The Wadi Al-Arab reservoir has been fed since 1986 in addition to the precipitation from the runoff of Wadi Al-Arab coming from the East Wadi Zahar from the South as well as partial pumping from the Yarmouk River. It has a storage capacity of 20×10^6 m of which 3.1×10^6 m^3 is dead storage 16.9×10^6 m^3 is available for irrigation of 3125 ha of land in the Jordan Valley (North Shuna). In drought periods this serves as drinking water supply for the capital Amman (Saadoun et al. 2010; personal communication H. Al-Shurieki dam director 2011). The estimated yearly discharge from the Wadi Al-Arab catchment with its 263.5 km^2 lays around 620×10^3 m^3 for the measured average discharge values from the hydrological year 1987–2012 (JDA data 2011). Pumping high evaporation rates have resulted in strong water level fluctuations throughout the years (Saadoun et al. 2008, 2010).

4.1 Field Campaign, Sampling and Laboratory Analyses

To evaluate the effects of the sediments on the storage as well as the current water quality, a field campaign was implemented in October 2011. The amount of water at this point in time was very low, with only 3.2×10^6 m^3 of water left in the reservoir. The lake had a surface extension of only 0.4 km^2 and the average water level was −127 m a.s.l. The focus of the campaign encompassed the following questions:

- What volume of sediment enters the reservoir each year and to what extent does this impact on the storage facility?
- How good is the water quality of the water reservoir? If there are contaminations, can sources be identified?

S. Kraushaar, *Soil Erosion and Sediment Flux in Northern Jordan*,
Springer Theses, DOI 10.1007/978-3-319-31888-2_4

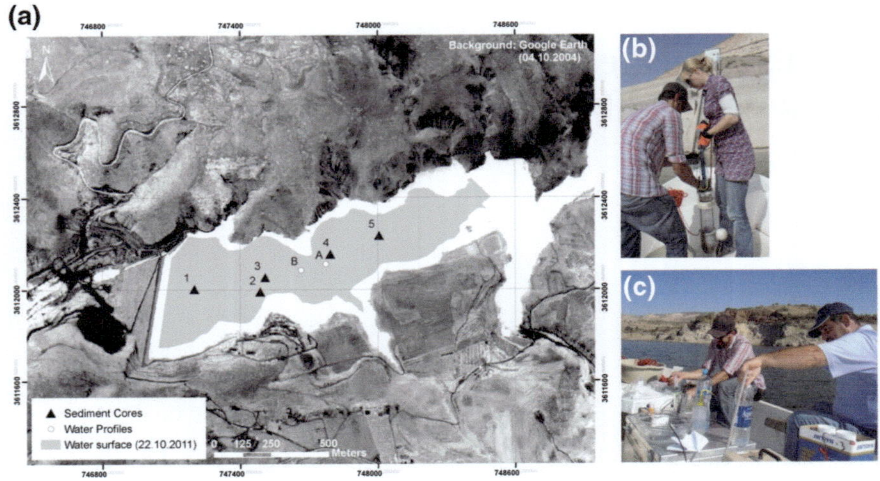

Fig. 4.1 Reservoir sampling in October 2011: **a** Sample Locations in Wadi Al-Arab Reservoir, **b** UWITEC Corer in action, and **c** physiochemical analysis and preparation of the samples (*Projection* UTM, WGS84, zone 36 N)

To answer these questions, the following samples/data were gathered (Fig. 4.1):

- Five sediment cores were taken from the bottom sediment with the UWITEC corer. The sediment cores showed a depth of up to 40 cm. All cores were sliced into 5 cm sample pieces, adding up to a total of 38 samples. Sampling locations were chosen so that the cores represented small as well as large scale variabilities in the lake.
- Two water profiles of 18 m (2 m steps) water column in the lake were investigated = 17 water samples.
- One echo sound campaign with the Lowrance LCX-15MT (50 kHz for areas <4 m depth; 200 kHz > 4 m depth) gave an additional 8000 relative measurement points of the water depth.
- A historical topographic map by the Japan International Cooperation Agency (JICA) pre dam building (1978) was digitalized.
    - The continuous interpolation of the existing data was implemented using the Triangular Irregular Network (TIN) method for the historical and Ordinary Kriging (trend removal 2) for the recent surface following the least mean square error when cross validating.
- Thirty-six background value samples were taken from the catchment's major geological units and land use units (Chap. 5):

- Eighteen geological rock samples were taken from the 3 major geological units: AHP/ASL, MCM, and URC.
- Eighteen soil samples derived from the 3 prevalent morphological positions with the respective land uses: olive orchards (slopes), agricultural fields (plains), and naturally vegetated slopes (slopes).

4.1.1 Sampling and Sample Preparation

The lake sediment samples were oven-dried for at least one week at 60 °C, carefully mortared, and sieved to 2 mm. The geological rock samples were crushed with a WC-coated Retsch BB0 jaw crusher to 2 mm size. Soil samples were air dried at 40 °C and sieved to 2 mm.

For the water samples, the water was pumped in 2 m steps from the selected depth (Fig. 4.1). The physiochemical parameters as redox potential, temperature, pH, and electric conductivity were analysed directly and in this order on the boat with a calibrated WTW Multiline P4. The order was important to guarantee minimal influence by oxygen exposure. The samples were filtered (≤0.45 μm CA-filter) and bottled (HDPE 60 ml) for further element analysis. Samples were stored at <5 °C and in darkness to reduce bioactivity. Bottles for cation analysis were acidified with 0.2 ml 10 % HCl to prevent the adsorption of the elements onto the walls of the bottle.

The point measurements were used to generate a bathymetric map of the recent sediment surface in the reservoir using Ordinary Kriging (trend removal 2). The historical digitalised topographic map was converted using the Triangular Irregular Network (TIN). Both methods were chosen following the least mean square error when cross validating. Using SURFER 9 (Golden Software), the resulting DEMs were subtracted to estimate the sedimentation volume.

4.1.2 Laboratory Analysis—Sediments

The grain size fractions were determined in the geochemical laboratory of TU Dresden following the Koehn method in accordance with E DIN ISO 11277:06.94. Due to the carbonatic background, an exception of the DIN was made concerning the destruction of organic carbon with hydrogen peroxide following Blume et al. (2002). Total inorganic content (TIC) and total organic content (TOC) of the samples were analysed with the LECO RC 412 Multiphase Carbon Determinator in accordance with DIN EN 15936.

Various elements as heavy metals, trace elements, as well as common cations and anions (Na, Mg, Al, Si, P, K, Ca, Ti, Cr, Mn, Fe, Ni, Cu, Zn, Br, Sr, Zr, Pb, Ba) were analysed with a sequential wavelength dispersive X-ray fluorescence analyser

(WDXRF–Bruker AXS S4 PIONEER) following the DIN EN ISO 12677 for sample preparation and an energy dispersive X-ray fluorescence analyser (EDXRF–SPECTRO X-LAB 2000 SPECTRO X-LAB 2000). Additionally, gamma ray emitting radionuclides of the ^{238}U- and the ^{232}Th decay chain as well as ^{40}K and ^{137}Cs were measured with two GAMMA-X, N-type coaxial high purity Geranium detectors (type GMX90-S; ORTEC). For further processing the software Gamma-W (Dr. Westmeier GmbH) was used.

Last but not least, 16 representative polycyclic aromatic hydrocarbons (PAHs) were analysed in selected catchment samples as well as in sediment samples from the bottom of the lake following (Murphy et al. 2012—Thermo Fischer Scientific). The PAHs emanate mainly from incomplete combustion processes as they take place while burning trash on the streets, as well as ashes, tyre debris, floods, and fertilisation and show a high sorption to soil particles and especially organic matter and are classified as carcinogenic (Guderian and Gunkel 2000).

4.1.3 Laboratory Analysis—Water

Major ions in the water samples were analysed with either ICP-MS (As, Cd, Cs, Fe, Mn, Rb, U: NIST Standard Reference Material 1643e) or ICP AES (P, B, Ba, Ca, Fe, K, Li, Na, Mg, Si, Sr: CIROS-Spectro Analytical Instruments by Kleve) calibrated with matrix adjusted standard solutions. NH_4 (DIN 38406-E5), PO_4 (DIN 38405-D11-1), NO_3 (DIN 38405-D10), and NO_2 (DIN 38405-D10) were photometrically determined using the EPOS Analyser 5060 (Eppendorf). Ion chromatography (ICS2000, Thermo Scientific–Dionex) following EN ISO 10304-2 and DIN 38405Cl was used for NO_3, SO_4, and Br.

Emerging compounds such as pharmaceuticals and personal care products (PPCPs) as well as chemicals used in agriculture (hexazinone, trifluraline, and tNP) were additionally analysed with a GC-MSD instrument (6890 HPMSD; Agilent Technology, Waldbronn, Germany) equipped with a 30 m HP-5MS (0.25 mm I.D., 0.25 μm film) following Eggen et al. (2010) and Möder et al. (2007). These analyses helped to estimate the anthropogenic effects on the deterioration of water quality.

4.2 Results and Discussion

4.2.1 General Physiochemical Features of the Wadi Al-Arab Reservoir

The Wadi Al-Arab reservoir shows monomictic characteristics as reported by Saadoun et al. (2010), where the mixing of the layers is related to the wet season

Fig. 4.2 Physicochemical parameters of the water profiles A and B

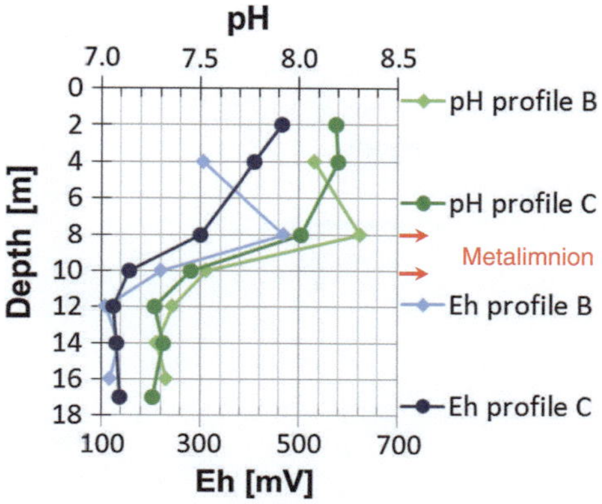

from autumn to spring. During the sampling in October 2011, the temperature and hence pH and the redox potential indicate the stratification of the water body.

The hypolimnion is 2.4 °C colder than the epilimnion and reaches down to a depth of 17 m to the reservoir sediments, with the metalimnion between 8 and 12 m deep (Fig. 4.2).

Here, the water profiles show an abrupt decrease from an alkaline (pH 8.2) to a neutral pH (pH 7.2). Additionally, below 8 m anoxic conditions prevail with Eh values <300 (Beutel et al. 2007). An increase of the electrical conductivity with depth might be related to these anoxic conditions that cause oxides such as iron and manganese oxides to become reduced and soluble (Beutel et al. 2007).

4.2.2 *Sedimentation in the Wadi Al-Arab Reservoir*

The sedimentation map in Fig. 4.3 shows areas of sedimentation and erosion, as well as areas that were hardly affected by sedimentation nor the construction of the dam. The sediment cores showed solely fine material <63 μm (in average 54 % silt and 46 % clay), which mainly settled to the west of the reservoir close to the dam wall, reaching a sediment body of 18 m height in the former depth of 35 m. The erosive areas are believed to be due to different reasons. These include the construction activities which took place between 1983 and 1986, which strongly affected the southern rim of the lake where a transport road was carved into the rock to serve the wall construction and the northern part, which was modified for the construction of the outlet and a spill over channel. Another possible reason might be a systematic error of predominantly the historical relief as the spatial topographic map by JICA is based only on 12 transects. However, subsurface debris cones connected to surface gullies indicate a good general fit of the map.

A total sediment volume of 1,735,505 m^3 was calculated, resulting in a yearly sedimentation volume of 66,750 m^3 ± 6 % since the flooding of the dam in 1986 corresponding to the estimation of around 70,000 m^3 $year^{-1}$ by the Japanese

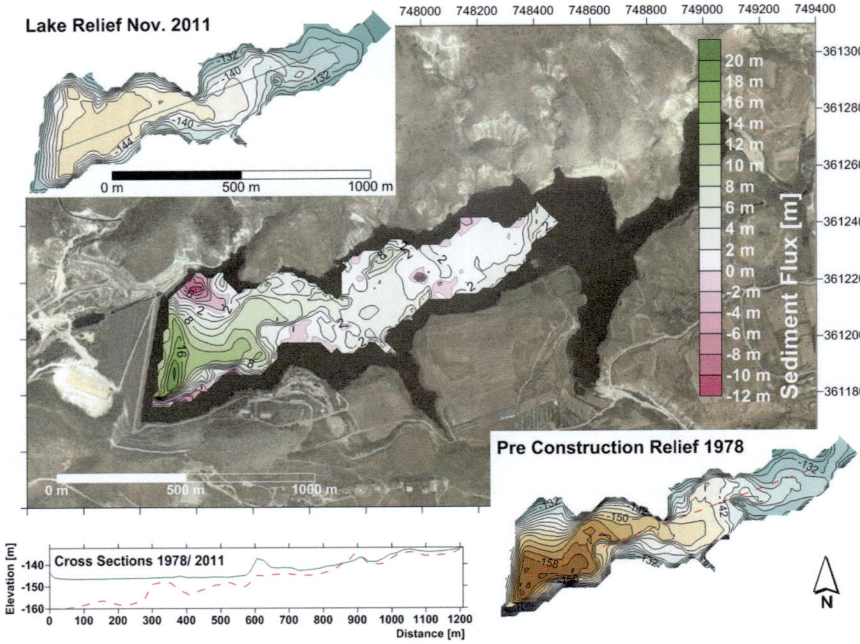

Fig. 4.3 Wadi Al-Arab sedimentation map with the lake relief of 1978 and 2011 (*Projection* UTM, WGS84, zone 36 N)

Fig. 4.4 Formerly deposited lake sediments (left upper corner) are being flushed into the smaller water body with the coming rain events (Kraushaar, 2011)

construction company (Nippon Koel and Nagai Eng. Co. 1978). This amount relates to about a 9 % loss of total storage volume in 26 years. This alone would not be a problem for the reservoir if the clayey sediments did not accumulate at the dam wall where the outlet is situated. Here, the sediment line is about 1–2 m below the outlet, posing an acute threat to the reservoir as there is no equipment available for the removal (personal communication H. Al-Shurieki, dam director, 2011). Furthermore, an acceleration of sedimentation can be postulated due to the low water tables in the last decade. At the end of summer when the water table is very low, high sediment packages (4–6 m) which have been flushed into the lake in former years are exposed at the rims of the reservoir. With the first rain events in September/October the sediments are then washed into the smaller surface of the reservoir, increasing the concentration of the sediment volume in the smaller water body (Fig. 4.4).

4.2.3 Sediment Quality in the Wadi Al-Arab Reservoir

The lake corers showed a layering of black and light grey sediments with a strong smell of hydrogen sulphide (H_2S), indicating the bacterial breakdown of organic matter in reduced oxygen conditions at the bottom of the lake. The analysis of the sediment samples showed no correlation between any of the analysed elements with the colouring of the layers. Hence, it seems reasonable that the black layers result due to the chemical reaction of iron and hydrogen sulphide to iron sulphide ($Fe + H_2S \rightarrow FeS + 2H^+$). Thereby, hydrogen ions are released, leading to a positive hydrogen budget in the hypolimnion and thus a reduced pH compared to the epilimnion (Fig. 4.2; personal communication Dr. Siebert, 2013).

The analysed element concentrations throughout the corer showed, in regard to the standard deviation, no differences in the profile, indicating that the sediment sources have not changed in the time considered. Therefore, the median element concentrations of the corer are used in the following qualitative assessment of some common (heavy) metal concentrations of the lake sediments. They were compared to the values from soil studies of Wadi Al-Arab (El-Radaideh 2010), soils of middle Europe (Blume 2004), the EU thresholds for the application of effluent sludge, and the USEPA (Unites States Environmental Protection Agency 2014) thresholds for (probable) effects on the environment and habitats. The red coloured median concentrations of the lake sediments exceed some, if not all, of the thresholds (Table 4.1).

Enrichments by a factor of two compared to the effluent sludge policy are visible in Chromium, Nickel, and Cadmium rendering the sediments as not suitable for soil replacement and amendment. Molybdenum shows higher values than other samples from the region or soils in Middle Europe; however no threshold effect level is identified for this element. Ghrefat and Yusuf (2006) also reported high cadmium contaminations in the sediments. For interpretation they compared the values to averaged worldwide geological background values by Ruppert (1991) and concluded that, due to the high amount, they must originate from the anthropogenic input of fertilizer and pesticide use.

In this study the geological background values sampled from the catchment were used for comparison, taking the matrix correction into account and using the Geo

Table 4.1 Selected element concentration of lake sediments compared to literature values and quality thresholds

Thresholds	mg kg^{-1}										
	Cr	Mn	Fe	Ni	Cu	Zn	As	Pb	Mo	Cd	Hg
Measured range	175	286	17910	78	46	204	7	9	5	2	<2
	256	348	23080	109	63	257	10	15	15	5	<2
Median	**228**	315.5	20035	**99.5**	55	227.5	8	23.5	**7**	**4**	<2
Wadi Al-Arab soils (Radaideh 2010)	n.d.	500–1000	n.d.	5–500	15–40	50–100	n.d.	10–67	0, 2–5	8	n.d.
Soils in middle Europe (Blume 2004)	n.d.	100–5000	1000–40000	n.d.	5–100	10–300	n.d.	n.d.	0, 3–5	n.d.	n.d.
EU policy for effluent sludge (Blume 2004)	30–100	n.d.	n.d.	30–50	50–140	150–300	n.d.	50–300	n.d.	1–3	1–1.5
Treshold effect level (USEPA 2014)	37.3	n.d.	n.d.	18	35.7	123	5.9	35	n.d.	0.60	0.17
Probable effect level (USEPA 2014)	90	n.d.	n.d.	36	197	315	17	91.3	n.d.	3.53	0.49

n.d. No Data
<2 = below Detection Limit

Accumulation Index (Eq. 4.1; Müller 1969) and the Enrichment Factor as described in Abrahim and Parker (2008; Eq. 4.2).

$$I_{geo} = \log_2 \frac{C_n}{B_n * 1.5} \tag{4.1}$$

where C_n is the measured parameter in the corer, B_n the measured geological background values, and 1.5 is related to the background matrix correction. Results rank the sediments in 7 enrichment classes, where an Index >5 means extremely contaminated and 0 uncontaminated.

$$EF = \frac{M_x * Fe_b}{M_b * Fe_x} \tag{4.2}$$

where M_x is the sediment sample concentration of any element, and Fe_x its iron concentration. Fe is used as one possible normalising element, acting as a "proxy" for the clay content alternative to the grain size fractions (Abrahim and Parker 2008). In this study, Al had a higher correlation with the clay fraction as Fe and therefore Al instead of Fe was used. M_b *and* Fe_b are the concentrations of the background material.

I_{geo} results show that Mn (3.6), K (3.3), Fe (3.1), Al (2.6), Ti (2.6), Zr (2.7), Rb (2.4), Si (2.0), and Ba (1.7) show strong (>3) to moderate (>1) contamination. The Enrichment Factor only identifies Mn (2) as enriched. Especially the elements Mn, Fe, Al, Ti, Zr, Rb, Si, and Ba are increased due to pedogenetic processes as solution weathering of the carbonatic geology and aeolian input (Sect. 2.4). These processes result in the augmentation of Mn, Fe, and Al as well as the trace elements (Lucke 2007; Blume et al. 2002). Potassium (K) is enriched probably due to algae growth in the reservoir. Hence, no anthropogenic impact can be retraced.

However, the results underpin the natural origin of the heavy metals chromium, nickel, and cadmium which exceed the qualitative thresholds (Table 4.1) in the geology. This results in the classification of the reservoir sediments in a geological contaminated site. Regarding the amount of clay, the high concentration of iron and the carbonatic background (mean 7.4 % TIC) and thus the pH above 7, the heavy metals are considered as immobile (Blume 2004). However, when hauled to the surface and used for agricultural use, plants can lower on small scale the pH and the heavy metals can become mobile (personal communication Prof. Dr. Vetterlein, 2013).

Concerning the radio nuclide contents, the sediment samples showed neither abnormalities nor worrying levels. Uranium (^{238}U) ranges between 100 and 200 Bq kg^{-1}, showing slightly lower values in the epilimnion as it is soluble in aerobic conditions. Potassium (^{40}K) ranges from 90 to 140 Bq kg^{-1} throughout the corer. Caesium 137 (^{137}Cs) as anthropogenic radio nuclide introduced into the atmosphere through nuclear bomb testing starting in the early fifties is, in regard to the measurement error, present in every sample ranging from 2.55 ± 0.39 to 15.87 ± 0.95 Bq kg^{-1}. This radio nuclide precipitates on the reservoir as well as the whole catchments and is bound to top soil particles, which then get transported to the reservoir. The concentration indicates that all sediments flushed into the reservoir once had surface contact. This leads to the assumption that subsurface erosion due to strong gullying is not the main erosion process in the catchment, as these sediments cannot contain ^{137}Cs. Furthermore, paleo terraces (containing no ^{137}Cs in the soil profile), which are present in the final Wadi stretches before the Wadi Al-Arab dam, must be geomorphological stable (Chap. 6).

The polycyclic aromatic hydrocarbons (PAHs) in the sediments have a summarised median value of 98.2 μg kg^{-1} throughout the corer, which is classified as unpolluted by the UNEP/WHO (1998). Only in the upper samples do concentrations range a bit higher (173 μg kg^{-1}). However, values below 1000 μg kg^{-1} are considered only as slightly contaminated and thus are not qualitatively alarming (Canadian Council of Ministers of the Environment 2010). Especially Benzo(a) pyren, which is considered as carcinogenic, is unproblematic with a value of <2 μg kg^{-1}.

4.2.4 *Water Quality in the Wadi Al-Arab Reservoir*

To detect the quality of the water ion analysis, an equivalent concentration check of the ion charges was implemented using the program PHREEQC (Version 3, USGS). Differences in the ion charges can result from ions that are bound in complexes, not measuring all ions, and inaccuracies during the measurements. Relating to a total equivalent concentration of 5 mmol l^{-1}, according to the German Working Group on water issues of the Federal States and the Federal Government (LAWA 1992) errors of up to 10 % are acceptable. From 13 water samples 9 showed error values below 10 %, 4 samples of profile B failed. All of these samples came from +12 m depths indicating higher error values of deeper water samples. Since both water profiles A and B showed similar developments, and all A profile samples showed good equivalent concentrations, these results are discussed in the following.

Regarding Fig. 4.5a, a high concentration of hydrogen carbonates (HCO_3), Na- and Ca-Ions, are visible which are influenced by the geological background of the region. The epilimnion shows enriched values of sulphates (SO_4), which are due to the aerobic decomposition of organic material. In the hypolimnion, organics are decomposed under anaerobic conditions, which lead to the production of hydrogen sulphide (H_2S) and consequently the typical smell of rotten eggs which was observed throughout the sampling on-site. The strong increase of ammonium together with the low nitrate values in the metalimnion might provide evidence for the high productivity of the water body (Fig. 4.5b; Schwoerbel 1999).

Concerning the tested heavy metals, only iron and manganese are pronounced in the hypolimnion (Fig. 4.5c). The increase close to the bottom sediments was firstly

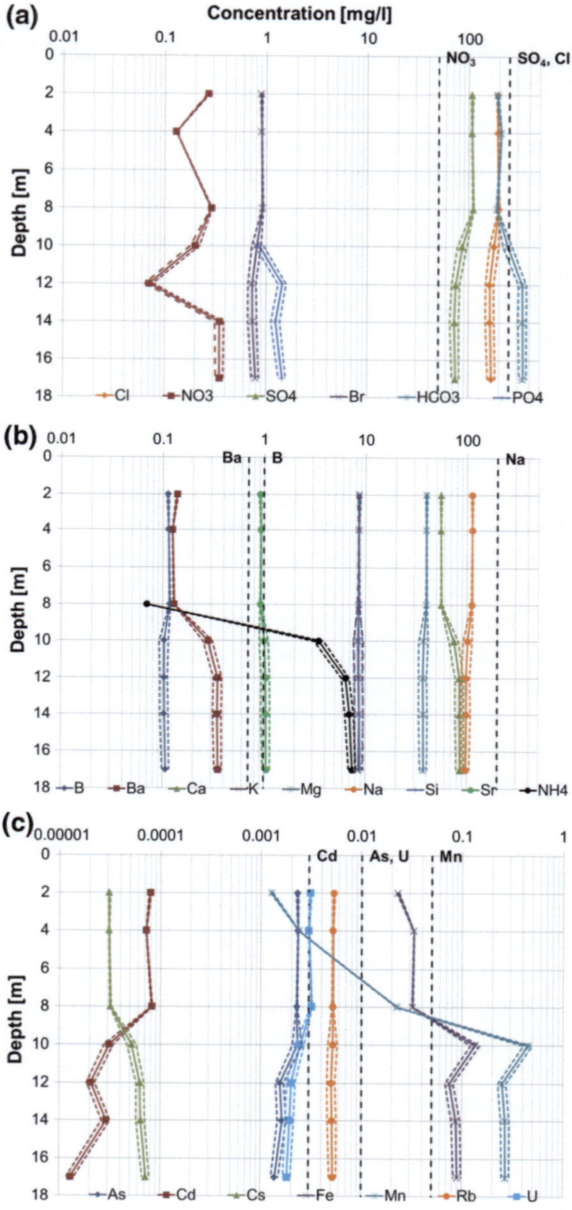

Fig. 4.5 Common anions (**a**), cations (**b**), and heavy metal (**c**) concentration of the water samples in Wadi Al-Arab reservoir (profile A). Black dashed lines = Quality thresholds of the TrinkwV 2001 / WHO 2011

interpreted as a leaching of the sediments. However, a comparison between the sediment ion concentrations and the water ion concentrations show—only in the case of Na—an increase of more than 1 %. For quality evaluation, the German Drinking Water Ordinance (TrinkwV 2001) and the international regulations by the WHO (2011) were consulted.

All physiochemical parameters stay within the limits, only manganese exceeds the threshold significantly from a depth of 10 m. The values of 0.16–0.45 mg l^{-1} in depths between 10 and 17 m surpass the threshold by a factor of 3–9.

Manganese is weakly toxic and causes negative effects on the body when repeatedly consumed. Although the iron concentration behaves similar to manganese, the threshold is not yet outrun. Obeidat et al. (2011) documented in a depth of 0.3 m a high concentration of copper (2.03 mg l^{-1}). This would be above the allowed maximum limit. Copper causes the water to taste metallic and is toxic for fish.

Further analysis of anthropogenic introduced compounds show that triclosan (disinfection and conservation, 1.152 ng l^{-1}), BPA (polymeric synthetics, 61.248 ng l^{-1}), trifluralin (herbicide, 0.128 ng l^{-1}), and caffeine (53.246 ng l^{-1}) exist in imperiling concentrations in the water (WHO 2011). Hexazinone (herbicide, 5.149 ng l^{-1}), tNP (fertilizer, 88.416 ng l^{-1}), galaxolide and tonalide (polycyclic musk aromas, 9.166 and 1.542 ng l^{-1}), and carbamazepine (anti-epileptic drugs; 21.12 ng l^{-1}) are proven to be present and harmful, too. However, no quality threshold exists for these compounds yet.

In conclusion, the water quality suffers mainly due to anthropogenic introduced compounds that find their way into the water body through effluent discharges of the two nearby waste water treatment plants east of the reservoir. Many times during fieldwork, and especially after rain events, discharge that originated from the treatment plants was observed and measured. Other possible sources for compounds such as caffeine, galaxolide, tonalide as well as triclosan and carbamazepine are damaged sewage tanks from the houses in the catchment and the result of illegal sewage spills by the sewage transport trucks.

As they are paid by load collected and not by load delivered, truck drivers were often witnessed disposing of the matter secretly in the Wadi instead of at the treatment plant. Fertilizers and herbicides such as trifluralin, hexazinone, and tNP are likely, however, to enter the water body bound to sediments that erode from agriculturally used areas and therefore document the contamination of the water resource related to erosion.

PAHs were not tested in the water samples. However, it has been documented that in anaerobic conditions they become soluble (Pravecek et al. 2005). As the general concentrations of the PAHs in the sediments were very low, no contamination related to them is assumed.

References

Abrahim GMS, Parker RJ (2008) Assessment of heavy metal enrichment factors and the degree of contamination in marine sediments from Tamaki Estuary, Auckland, New Zealand. Environ Monit Assess 136, 227–238

Beutel MW, Leonarda TM, Denta SR, Mooreb BC (2007) Effects of aerobic and anaerobic conditions on P, N, Fe, Mn, and Hg accumulation in waters overlaying profundal sediments of an oligo-mesotrophic lake. Water Res 42:1953–1962

Blume HP (ed) (2004) Handbuch des Bodenschutzes. Bodenökologie und-belastung—Vorbeugende und abwehrende Schutzmaßnahmen, 3rd edn. Ecomed, Landsberg am Lech, pp 723–769

Blume H-P, Brümmer GW, Horn R, Kandeler E, Kögel-Knabner I, Kretzschmar R, Stahr K, Wilke B-M (2002) Scheffer/Schachtschabel: Lehrbuch der Bodenkunde, 15th edn. Spektrum Akademischer Verlag GmbH, Heidelberg/Berlin

Canadian Council of Ministers of the Environment (2010) Canadian soil quality guidelines. Carcinogenic and other polycyclic aromatic hydrocarbons (PAHs). (Environmental and Human Health Effects). Scientific Criteria Document, Quebec

Eggen T, Moeder M, Arukwe A (2010) Municipal landfill leachates: a significant source for new and emerging pollutants. Sci Total Environ 408:5147–5157

El-Radaideh NM (2010) Using bottom reservoir sediments as a source of agricultural soil: Wadi El-Arab reservoir as a case study, NW Jordan. Abhath Al-Yarmouk: Basic Sci Eng 19(2):75–91

Ghrefat H, Yusuf N (2006) Assessing Mn, Fe, Cu, Zn, and Cd pollution in bottom sediments of Wadi Al-Arab Dam, Jordan. Chemosphere 65:2114–2121

Guderian R, Gunkel G (2000) Handbuch der Umweltveränderungen und Ökotoxikologie. Band 3B: Aquatische Systeme: Biogene Belastungsfaktoren—Organische Stoffeinträge—Verhalten von Xenobiotika. Springer Verlag, Berlin/Heidelberg

Lucke B (2007) Demise of the Decapolis. Past and present desertification in the context of soil development, land use and climate. Dissertation at the Brandenburg Technical University of Cottbus

Möder M, Braun P, Lange F, Schrader S, Lorenz W (2007) Highly selective and sensitive determination of endocrine disrupting compounds and pharmaceutical residues using solid phase extraction, derivatization and gas chromatography—negative chemical ionization mass spectrometry. Clean 35:441–451

Müller G (1969). Index of geoaccumulation in sediments of the Rhine River. Geojournal 2:108–118

Murphy B, Lingam S, Richter B, Carlson R (2012) Simultaneous extraction of PAHs and PCBs from environmental samples using accelerated solvent extraction. Thermo Fisher Scientific, Application Note 1025, Salt Lake City, USA

Nippon Koel and Nagai Eng. Co. (1978) Detailed Survey and Investigation Report on Wadi Arab dam and Irrigation Project

Obeidat SM, Alomary A, Sekhaneh W, Al-Momani I, Al Abdel Hamid AJ (2011) Assessment of water quality in four main water reservoirs in Northern Jordan. Int J Chem 3(2):79–87

Pravecek TL, Christman RF, Pfaender FK (2005) Impact of imposed anaerobic conditions and microbial activity on aqueous-phase solubility of polycyclic aromatic hydrocarbons from soil. Environ Toxicol Chem 24(2):286–293

Ruppert H (1991) Natürliche Spurenmetallgehalte im Boden und ihre anthropogene Überprägung. Mitt Österr Geol Ges 83:243–265

Saadoun I, Bataineh E, Al-Handal AY (2008) The primary production conditions of Wadi Al-Arab Dam (reservoir), Jordan. Jordan J Biol Sci 1(2):67–72

Saadoun I, Bataineh E, Al-Handal A, Hindeh M (2010) Physicochemical features of Wadi Al-Arab Dam (reservoir), Jordan. Int J Oceanogr Hydrobiol 39(4):191–205

Schwoerbel J (1999) Einführung in die Limnologie, 8th edn. G. Fischer, Stuttgar

Trinkwasserverordnung (TrinkwV) (2001) http://www.gesetze-im-internet.de/trinkwv_2001/index.html (last checked 05.05.2014)

United Nations Environment Programme, International Labour Organisation, World Health Organisation (UNEP/ WHO) (1998) International Program on Chemical Safety. Environmental Health Criteria 202. Selected non-heterocyclic polycyclic aromatic hydrocarbons

United States Environment Proteciton Agency (USEPA) (2014) Water quality standards handbook—Chapter 3: Water Quality Criteria (40 CFR 131.11) http://water.epa.gov/scitech/swguidance/standards/handbook/chapter03.cfm (last checked 12.06.2014)

Working Group on water issues of the Federal States and the Federal Government (LAWA) (1992) Grundwasserrichtlinie, Teil 3: Grundwasserbeschaffenheit

World Health Organization (WHO) (2011) Guidelines for Drinking-Water Quality, 4th edn. WHO Press, Geneva

Chapter 5
Sediment Fingerprinting: A Revised Approach for Data Correction and Evaluation

The information in this Chapter are published by Kraushaar et al. (2015) *in Journal of Soils and Sediments 15, 2155–2173.*

5.1 Introduction

In semi-arid regions, where water is scarce, soil erosion on slopes decreases the water infiltration potential, causes accelerated surface water runoff, and thus reduces the recharge of aquifers. Eroded sediments are flushed into surface water reservoirs and often deteriorate the water quality and the reservoirs' capacity (Palmieri et al. 2001; Slimane et al. 2013; Vanmaercke et al. 2011). In the extremely water-scarce nation of Jordan, surface reservoirs play a tremendous role in freshwater supply (Nortcliff et al. 2008). Increased sediment deposition due to erosion processes was reported by several authors in water reservoirs along the Lower Jordan Valley as they are the final sink for surface runoff and sediments (Ghrefat and Yusuf 2006; El-Radaideh 2010; Al-Ansari and Shatnawi 2011; Al-Ansari et al. 2012). To mitigate this problem, it is of great importance to define relevant sediment sources and their respective contribution to the sink.

Sediment fingerprinting has been proved to be a valuable tool for this task (Davis and Fox 2009; Owens and Xu 2011; Mukundan et al. 2012; Koiter et al. 2013a). Each potential source is characterized by discriminating properties or elements. By applying a multivariate mixing model, the source composition is iteratively determined by minimizing the sum of squared weighted errors of the element concentrations between the estimated source mixture and sink (Yu and Oldfield 1989; Walling et al. 1993; Collins et al. 2010; Wilkinson et al. 2012).

Electronic supplementary material The online version of this article (doi:10.1007/978-3-319-31888-2_5) contains supplementary material, which is available to authorized users.

S. Kraushaar, *Soil Erosion and Sediment Flux in Northern Jordan*,
Springer Theses, DOI 10.1007/978-3-319-31888-2_5

A variety of discriminating properties are used, for example, grain size, sediment colour, plant pollen, mineral magnetism, organic and inorganic elements, radionuclides, stable isotopes, rare earth elements and biogenic properties (Foster and Lees 2000; Collins and Walling 2002; Walling 2005, 2013; Davis and Fox 2009; Mukundan et al. 2010). The basic assumption of the method is that suitable properties are comparable between source and sink, and that they stay conservative during transport (Davis and Fox 2009; Koiter et al. 2013a). However, particle selection during transport can lead to an accumulation of smaller particles and organic matter with a higher specific surface area (SSA) in the sink (Koiter et al. 2013a; Walling 2013). Moreover, the larger the SSA of a particle, the more the elements are adsorbed to it and thus have the potential to hinder a direct elemental comparison between source and sink (Horowitz 1991; Collins et al. 1997; Motha 2002; Koiter et al. 2013a; Walling 2013). In contrast, high carbonate concentrations in a sample dilute the elemental content compared to those with lower carbonate concentrations (Horowitz 1991).

To make a successful comparison between source and sink, one may restrict the element analysis to specific grain size fractions, for example, <2 to <63 μm (Motha et al. 2004; Collins et al. 2012; Wilkinson et al. 2012). This requires profound knowledge of the transport and the accompanied change in grain size composition (Koiter et al. 2013a, b; Walling 2013). Another option is to use correction factors (Collins et al. 1997, 2010, 2012; Walling et al. 1999; Russell et al. 2001; Krause et al. 2003; Foster et al. 2007; Koiter et al. 2013a; Smith and Blake 2014). Some authors additionally remove organic carbon as a second correction (Motha 2004; Wilkinson 2012).

Since the study presented here was preliminary, no previous knowledge existed concerning grain size fractions, particle selection, organic/inorganic carbon contents and their effects on the element concentrations. Consequently, we decided to apply correction factors to account for the comparability of the sediments, since selective grain size sampling or the removal of organic matter would have implied the possibility of overlooking important information within the data.

Correction factors were introduced by Collins et al. (1997) and have become a widely used procedure in fingerprinting studies (Walling et al. 1999; Krause et al. 2003; Foster et al. 2007; Collins et al. 2010, 2012). Correction factors have to be determined for those parameters that influence the element concentration, such as grain size and/or total organic carbon (TOC). To derive grain size correction factors, dividing the SSA in sink samples by the average SSA of the source samples has become a widely accepted approach (Walling et al. 1999; Krause et al. 2003; Foster et al. 2007; Collins et al. 2010, 2012). To correct the effects of TOC, the factor-determination is similar and the TOC content of the sink is divided by the TOC content of the source (Collins et al. 1997). In the mixing model, both correction factors are sometimes used as multipliers for each mean source element concentration, assuming that both influence each element independently but in the same positive linear way.

However, Collins et al. (1997) and Mukundan (2012) report that the variances in element concentrations resulting from the influences of particle size and TOC often overlap. Thus, interrelations between the influential parameters particle size and TOC exist and correcting for both would make an overcorrection likely. Russell et al. (2001) found that elements were positively correlated with logarithmic SSA. Furthermore, they argued that the TOC influence is negligible compared to the particle size and is due to the interrelations already covered by the particle size correction. Smith and Blake (2014) report that SSA and TOC may vary strongly due to temporal changes in and between the sources and not exclusively due to particle selectivity. Furthermore, they found inconsistencies of direction and strength in their linear correlation analysis between geochemical elements and SSA and TOC. They concluded that establishing specific relationships between elements and grain size fractions and correcting only for SSA should be done solely in cases where correction for particle selectivity seems appropriate (Smith and Blake 2014).

Further difficulties appear if changing conditions overturn the conservative behaviour of selected properties. This is the case in the study area, where sediments get flushed from an oxic environment into the Wadi Al-Arab Dam, where they settle into an anoxic environment. There, pH changes and important bio-geochemical reactions can occur, particularly on sediment–water interfaces (Stumm and Morgan 1996; Koiter et al. 2013a), possibly rendering fingerprinting elements non-conservative (Owens and Xu 2011; Koiter et al. 2013a). However, although well known, the influence of changing environments on, for example, dissolution or complexation of elements has rarely been discussed in fingerprinting studies (Salomons and Mook 1987; Horowitz 1991; Kelley and Nater 2000; Miller et al. 2005; Beutel et al. 2007; Foster et al. 2007; Kouhpeima et al. 2010; Slimane et al. 2013).

The objectives of the present study were: (1) to introduce an alternative estimation of element specific correction factors based on linear and non-linear multiple regression analysis; (2) to include physicochemical water parameters in the selection process; (3) to evaluate the conservative nature of the fingerprinting elements; and (4) to calculate the relative contribution of each sediment source to the Wadi Al-Arab reservoir.

5.2 Study Area

The Wadi Al-Arab catchment extends over 263.5 km^2 in the north west of Jordan. The Wadi itself cuts its way from the Jordan Valley Escarpment (500 m a.s.l.) down to the Jordan Valley (−164 m a.s.l.) and is a tributary to the Jordan River. The annual precipitation ranges from ~530 mm in the east to ~380 mm in the west, resulting in a Mediterranean to semi-arid climatic regime. The catchment's bedrock mainly consists of limestone and marl with terrestrial input (Al-Sharhan and Nairn 1997; Moh'd 2000). Figure 5.1 displays the extension of the three major lithological units of the Upper Cretaceous and Paleocene with the oldest unit (Al-Hisa Phosphorite/Amman Silicified Limestone) in the east and the youngest (Umm Rijam Chert) in the north and northwest.

Typical soil types for the region are Calcaric Cambi- or Lithosols on the slopes and Chromic Vertisols or Vertic Cambisols on top or saddle positions (MoA 1994). A plateau used for agriculture characterizes the east of the catchment. From there,

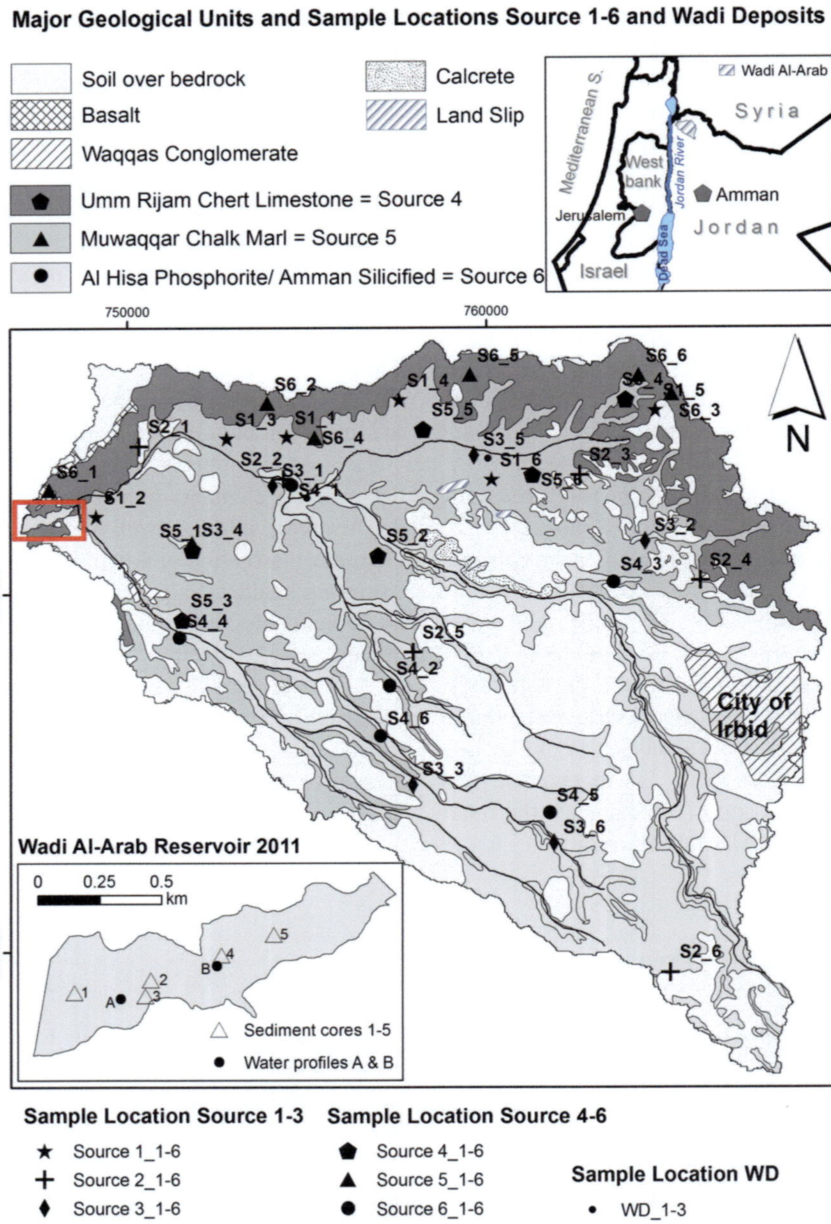

Fig. 5.1 Major geological units and sample locations in Wadi Al-Arab, Jordan

an increase in relief is observed to the west and south. Bare rock/soil (34 %), olive orchards (27 %), fields (14 %), and natural shrubs on steep slopes (11 %) are the major land cover units in the catchment. At the western outlet, the Wadi Al-Arab reservoir collects the overland flow for agricultural projects in the Jordan Valley as well as, during periods of water shortage, for households in the capital, Amman. From our own bathymetric investigations in October 2011 and from GIS analysis, a yearly sedimentation volume of 67,000 $m^3 \pm 6$ % was calculated for the Wadi Al-Arab lake, posing an acute threat to the reservoir due to the position of the dam's outlet (Hussein Al-Shurieki 2011, personal communication, dam director; unpublished data 2011).

5.2.1 *Definition of Sediment Sources*

The erosive potential of several land units was the subject of previous work. Under olive trees on slopes, splash, rill, and gully erosion occur and follow the downhill furrows, channelising the runoff. Stone pedestals and siltation features in the rills and gullies are also visible and tillage erosion is likely (Kraushaar et al. 2014). On agricultural fields, ephemeral gullies as well as piping were occasionally observed. Naturally vegetated slopes experience intensive grazing and thus reduced resistance to water erosion (Yassoglou and Kosmas 1997; unpublished results). In areas with sparse or no vegetation cover, such as south-exposed slopes or accumulations of new sediments from road construction, the accretion of rills, stone pedestals, gullies as well as the occurrence of badlands are common. The study aims to differentiate between these main sources: olive orchards (Source 1 = S1; Fig. 5.2a); fields for agriculture (Source 2 = S2; Fig. 5.2b); naturally vegetated slopes (Source 3 = S3; Fig. 5.2c); and the unconsolidated rock and recent sediments of the three main geological units Umm Rijam Chert (Source 4 = S4; Fig. 5.2d), Muwaqqar Chalk Marl (Source 5 = S5; Fig. 5.2e), and Al-Hisa Phosphorites/Amman Silicified Limestone (Source 6 = S6; Fig. 5.2f).

It was known from the literature that the heavy metal concentrations of the three different geological units were significantly different (Al-Sharhan and Nairn 1997; Moh'd 2000; Moh'd and Powell 2010). Furthermore, observations showed that the soil development on slopes (geomorphologically active) compared to the top and saddle positions (geomorphologically stable) is marginal. This shows in the enrichment of conservative elements due to solution weathering of the carbonatic rock and pedogenesis (Lucke 2007). The use of fertilizers and other agricultural additives is reflected in certain geochemical elements, for example, N, P, K and Cd (Ghrefat and Yusuf 2006). Finally, regular tillage practice is known to homogenize the concentration of fallout radionuclides, such as ^{137}Cs, to the depth of the plough layer. Based on these findings, a differentiation between the mentioned sources seemed possible. Sources S1–S3 represent surface erosion from the mentioned land use units. Sources S4–S6 represent subsurface erosion from linear features, as well as unconsolidated rock and new sediments, differentiated by the three main geological units (Table 5.1; Fig. 5.2).

Table 5.1 Characteristics and potential of discriminating elements for the six defined sediment sources

Source no.	Morphology	Soil type	Land use	Processes assumed to generate characteristic element contents	Number of samples
S1	Steep slopes >3–4°	Shallow yellow med. soils	Cultivated mainly with olives, ploughed 2–4 times per year	– ^{137}Cs and $^{210}Pb_{ex}$ ploughed into surface layer-> concentration less intense – Enrichments of fertilizer in top soil (NPK) – Medium content of pedogenetic enriched elements such as Fe, Ti, Al	6
S2	Stable positions <3° (plateau and saddle pos., terraces)	Deep red med. soils	Cultivated mainly with wheat and vegetables, ploughed >4 times per year	– ^{137}Cs and $^{210}Pb_{ex}$ ploughed more homogenously into surface layer-concentration less intense – Conservative elements (heavy metals, Mn, Al, Fe) enrich with weathering – Enrichments of fertilizer in top soil (N, P, K)	6
S3	Steep slopes >3–4°	Shallow yellow med. soils	Not cultivated, natural vegetation, grazing	– ^{137}Cs and $^{210}Pb_{ex}$ concentrated on the surface	6

Source no.	Morphology	Sediment type	Geolgical unit	Processes assumed to generate characteristic element contents	Number of samples
S4	On slopes as well as in leveled positions	Sub surface samples of (unconsolidated) rock representing: – initial (new) sediments –subsurface erosion -> gullies, C-horizons	Umm Rijam Chert (URC)	– No ^{137}Cs, no $^{210}Pb_{ex}$	6
S5			Muwaqqar Chalk Marl (MCM)	– Less pedogenetic enriched and conservative elements – High Ca content – Differences in the genesis lead to differences in the element content	6
S6			Al-Hisa Phosphorites/Amman silicified limestone (AHP/ASL)	– No ^{137}Cs, no $^{210}Pb_{ex}$ – Less pedogenetic enriched and conservative elements – High Ca content – Higher uranium content of phosphorites	6

Fig. 5.2 Photographs of the six sediment sources: **a** sloped cultivated olive orchards (S1), **b** agricultural fields on plateau and saddle positions (S2), **c** slopes used for grazing (S3), **d** Umm Rijam Chert (S4), **e** Muwaqqar Chalk Marl (S5), and **f** Amman Al-Hisa Phosphorites/Amman Silicified Limestone (S6)

5.3 Methods

5.3.1 Source Sampling

For sources S1–S3, six composite soil samples were obtained from different locations throughout the catchment (Fig. 5.1). Each composite sample consisted of 10–20 subsamples depending on the size of the orchard, field or slope and covered areas of 0.5–1 ha. Subsamples were collected along transects of 15–20 m in width, at a depth between 0–5 cm, and sieved to <2 mm. About 1 kg of soil was collected in this way per site, focusing on areas and material that are likely to erode, such as aggregates in downhill furrows or loose soil on bare rock on a slope.

Sources S4–S6 represent subsurface samples and thus included consolidated and unconsolidated geology. Consolidated geology samples were collected with a hammer and chisel from exposed rocks. Care was taken to ensure that the weathering crust and no material that had direct surface exposure entered the sample. In case of unconsolidated material, such as the Muwaqqar Chalk Marl (Fig. 5.1), 10 sub samples were collected with a plastic shovel as a composite sample from a depth of >30 cm.

Additionally, three samples of alluvial Wadi deposits (WD) along the principal valley of Wadi Al-Arab were collected. Sampling sites included two alluvial terraces from which material was derived from the cleaned profiles at a depth of >50 cm and one recent alluvial deposit, from which a surface sample of 0–5 cm depth was taken. The aim of these samples was to gather information on the composition of grain size fraction during transport, and included the element concentration in the modified element correction approach.

5.3.2 Sink Sampling

5.3.2.1 Wadi Al-Arab Dam Reservoir Sediments

Five reservoir sediment cores with an average length of 40 cm were collected with a gravity corer of UWITEC at different locations of the reservoir (Fig. 5.1). Each sediment core was sliced into 5 cm intervals resulting in a total of 38 samples. This sampling strategy was chosen to represent the spatial and vertical variability of potential fingerprinting elements in the lake sediment.

5.3.2.2 Wadi Al-Arab Dam Reservoir Water Profiles

In addition, two profiles over the whole water column (18 m) were investigated to compare water element concentrations with those of the reservoir sediments. A peristaltic pump was used to sample water in 2 m steps from the air–water surface down to the bottom sediment (n = 17). Physiochemical parameters (Eh, T, pH, EC) were analysed directly with a calibrated WTW Multiline P4. The measurement of the redox potential was done to avoid any significant influence of air oxygen on the measured value. Each sample was filtered (0.45 μm CA-filter) and filled into two HDPE bottles (60 ml) for cation and anion analysis. Cation samples were acidified with 0.2 ml 10 % HCl to prevent adsorption on the sample bottle. The samples were kept cool (<5 °C) and dark until analysis.

5.3.3 Laboratory Analysis

For sample preparation, the reservoir sediment samples were oven dried at 60 °C until constant weight. The resulting aggregates were carefully ground with a mortar and pestle for further analysis. The consolidated rock samples were crushed to a particle size of 2 mm with a jaw crusher (Retsch type B00). Table 5.2 gives an overview of the different analyses and instruments used on the sediment samples.

The grain size fractions were determined in the geochemical laboratory of the Technical University of Dresden following the Koehn method according to E DIN ISO 11277:06.94. Due to the generally low TOC content (mean: 2 %) compared to the higher inorganic carbon content (mean: 6.5 %), an exemption of the DIN was made concerning the destruction of organic carbon with hydrogen peroxide following Blume et al. (2004). The elimination of organic carbon using H_2O_2 on heat plates would lead to the destruction of the carbonatic sand particles into silt and clay and cause a high error in these fractions (Döpke 2004). Thus, the TOC was not removed before the grain size analysis and therefore appears as an average error of 2 %, mainly in the sand fraction. Total inorganic content (TIC) and TOC of the

Table 5.2 Overview of samples, analyses, and instruments

Sample ID	Number of samples	Grain size[a]	Element concentration[b,e]	Radionuclides[c]	TOC/TIC contents[d]
Source 1	6	X	X[b]	X	X
Source 2	6	X	X[b]	X	X
Source 3	6	X	X[b]	X	X
Source 4	6		X[b]	X	X
Source 5	6		X[b]	X	X
Source 6	6		X[b]	X	X
WD	3	X	X[b]	X	X
Core 1	6		X[b]	X	
Core 2	8		X[b]	X	X
Core 3	8		X[b]	X	
Core 4	8	X			
Core 5	8		X[b]	X	X
Water profile A	6		X[e]		
Water profile B	7		X[e]		

X = Analyses was performed

[a]Pipette method following Köhn (DIN ISO 11277: 08.02)

[b]X-ray fluorescence (WDXRF, AXS S4 PIONEER, Bruker: DIN EN ISO 12677): Na, Mg, Al, Si, P, S, K, Ca. Energy dispersive X-ray fluorescence analysis (EDXRF, X-LAB 2000, SPECTRO): Ti, V, Cr, Mn, Fe, Co, Ni, Cu, Zn, Se, As, Br, Rb, Sr, Zr, Pb, U, Th, Mo, Cd, Hg, Sn, Sb, Ba

[c]N-type coaxial HPGe gammaspectrometer (GAMMA-X GMX90-S, ORTEC): ^{238}U, ^{226}Ra, ^{210}Pb, ^{232}Th, ^{228}Th, ^{40}K, ^{137}Cs

[d]Infrared absorptions photometry (RC 412 Multiphase Carbon Determinator, LECO: DIN EN 15936): TIC, TOC

[e]Inductively coupled plasma mass spectrometry (ICP-MS, Elan DRCe, Perkin-Elmer): As, Cd, Cs, Fe, Mn, Rb, U. Optical emission spectrometry ICP-OES (Spectro "Ciros", Kleve): B, Ba, Ca, K, Li, Mg, Na, P, Si, Sr. Ion chromatography (ICS2000, Thermo Scientific (Dionex): EN ISO 10304-2, DIN 38405): Cl, NO_3, SO_4, Br. Photometry (EPOS Analyser 5060, Eppendorf: DIN 38406/05): NH_4, PO_4, NO_3, NO

samples were analysed with a LECO RC 412 Multiphase Carbon Determinator in accordance with DIN EN 15936.

An energy dispersive X-ray fluorescence analyser (EDXRF—SPECTRO X-LAB 2000 SPECTRO X-LAB 2000) was used to determine the elements As, Ba, Br, Cr, Cu, Fe, Mn, Ni, Pb, Rb, Sr, Th, Ti, U, V, Zn, and Zr. For sample preparation, the homogenized sample mixture (4 g) and wax (0.9 g) was pressed to a pellet. Elements such as Al, Ca, K, Mg, Na, P, S, and Si of a sample were analysed with a sequential wavelength dispersive X-ray fluorescence analyser (WDXRF—Bruker AXS S4 PIONEER) following the DIN EN ISO 12677 for sample preparation. A calibration function was generated for both XRF instruments with standards of a known sample with similar matrix.

Additionally, gamma ray emitting radionuclides of the ^{238}U- and the ^{232}Th decay chain as well as ^{40}K and ^{137}Cs were measured with two GAMMA-X, N-type coaxial high purity Geranium detectors (type GMX90-S; ORTEC). IAEA standards of similar matrix to the samples were used for the calibration of the detector. The sample material was filled in cylindrical plastic containers with a volume of 108 cm^3 or 32 cm^3, weighed and air-sealed to avoid the escape of the ^{222}Rn gas and keep the activity equilibrium between the radionuclides of the ^{238}U decay. The measurements were undertaken three weeks after filling to guarantee the activity equilibrium between long-life and short-life radionuclides. Measurements were performed for 12 h per sample; further processing was done using the Gamma-W software (Dr. Westmeier GmbH). $^{210}Pb_{ex}$ was calculated by subtracting the specific activity of ^{226}Ra from the specific activity of ^{210}Pb.

Major ions in the water samples were analysed with either ICP-MS (As, Cd, Cs, Fe, Mn, Rb, U: NIST Standard Reference Material 1643e) or ICP AES (P, B, Ba, Ca, Fe, K, Li, Na, Mg, Si, Sr: CIROS-Spectro Analytical Instruments by Kleve) calibrated with matrix adjusted standard solutions. NH_4 (DIN 38406—E5), PO_4 (DIN 38405—D11-1), NO_3 (DIN 38405—D10), and NO_2 (DIN 38405—D10) were photometrically determined using the EPOS Analyser 5060 (Eppendorf). Ion chromatography (ICS2000, Thermo Scientific—Dionex) following EN ISO 10304-2 and DIN 38405Cl was used for NO_3, SO_4 and Br.

5.3.4 Data Correction—Revised Data Correction Factor

In the carbonatic catchment of Wadi Al-Arab, we considered the sand, silt, and clay content, the total organic carbonate as well as the TIC percentage to potentially influence element concentrations, as described by Horowitz (1991). These parameters that affect the element content of a sample will be labeled as influential parameters. A stepwise multiple regression analysis model (SMRAM) was implemented to investigate linear and non-linear relationships, according to the best fit, between grain size fractions, TOC and TIC, and the geochemical elements and radionuclides, respectively. Results were incorporated into an element correction factor while taking account of possible interrelations between all the influencing parameters.

The basic assumption of the method is that a part of the concentration variance of an element i ($element_i$) can be explained by the concentration variances of different influential parameters. The common variances need to be minimized to reliably relate source to sink samples. Applying a regression model, the resulting R^2 indicates the impact, and the slope of the regression represents the intensity of the influential parameter on the element concentration. However, several influential parameters can impact the variance of the $element_i$ while at the same time also impacting each other (inter-correlation). Establishing correction factors independently for the parameters could lead to overcorrection. Therefore:

1. The SMRAM is driven on semi partial coefficients and Eq. (5.1) was chosen to simulate the natural concentration variance of each element $(\widehat{Y}_i)$.

2. An element and source specific correction factor (CF_{is}) is produced from the ratio between simulated sink $(\widehat{Y}_{i_{sink}})$ and mean simulated source concentration of element$_i$ $(\widehat{Y}_{is})$ (Eq. 5.2).
3. Finally, a corrected element data set Yc_{is} is generated for the mixing model with Eq. (5.3).

The example given in Fig. 5.3 starts with bi-plots of the raw data of each element $(Y_{raw}$; n $= 21)$ in relation to the influencing parameters. Here, the two with the highest R^2 values at p-level = 0.05 (*) are TIC and clay (Fig. 5.3a, b). The residuals of the regression with the highest R^2 (Fig. 5.3a) are then taken to a semi-partial regression analysis (Fig. 5.3c) to explore the interrelations between TIC and clay. In the case of the example, the best fit was given by a linear relationship. However, for each element different positive and negative, linear and non-linear models were appropriate and therefore used. The semi-partial regression analysis is repeated until no more significant R^2 values between the residuals and the influencing parameters occur. In the example of Fig. 5.3, Si only needs to be corrected for TIC.

Once all significant influential parameters are identified for each element and their influences quantified in regression equations, they are summarised in one element-specific multivariate equation to merge the effects:

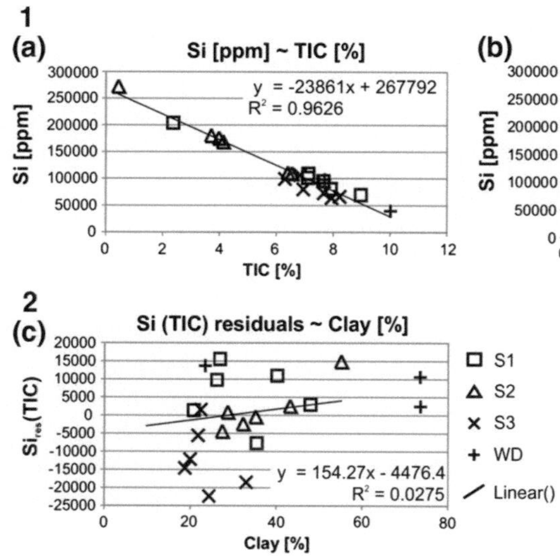

Fig. 5.3 The SMRAM model implementation in two steps: (1) regression analyses between Si and **a** influential parameter 1 (= TIC), **b** influential parameter 2 (= Clay). (2) Regression **a** shows a higher R^2 and therefore the residuals of Si (TIC) are then plotted against influential parameters 2 (= Clay) in a semi partial regression analyses. The low R^2 of graph **c** indicates that the variance of TIC also explained the variance of clay

$$\widehat{Y}_i = Y_{raw}(IP_1) + Y_{resIP_1}(IP_2) + \cdots + Y_{resIP_{n-1}}(IP_n) \quad (5.1)$$

where: $\widehat{Y}_i$ = simulated concentration of element$_i$ based on the related influential parameters; $Y_{raw}(IP_1)$ = regression equation of Y_{raw} and influential parameters$_1$ with highest R^2; $Y_{resIP_1}(IP_2)$ = semi-partial regression equation with the residual of $Y_{raw}(IP_1)$ and influential parameters$_2$ with highest R^2; and n = number of influential parameters.

Using Eq. (5.1) the mean element concentration of the various sources and the sink is simulated based on the quantified relation of the relevant influential parameters. This allows one to relate the simulated concentrations of sink $(\widehat{Y}_{i_{sink}})$ and source $(\widehat{Y}_{is})$ and defines the source and element-specific correction factors (CF_{is}):

$$CF_{is} = \frac{\widehat{Y}_{i_{sink}}}{\widehat{Y}_{is}} \quad (5.2)$$

where:

CF_{is} correction factor of element$_i$ and source$_s$;
$\widehat{Y}_{i_{sink}}$ simulated mean sink concentration of element$_i$;
$\widehat{Y}_{is}$ simulated mean concentration of element$_i$ in source$_s$.

The correction factors are then used to correct the raw element concentrations of the sources:

$$Yc_{is} = r_i * CF_{is} * Yraw_{is} + (1 - r_i) * Yraw_{is} \quad (5.3)$$

where:

Yc_{is} corrected concentration of element$_i$ in source$_s$;
r_i adjusted R^2 value of the multivariate regression equation;
CF_{is} corresponding correction factor of element$_i$ in source$_s$; and
$Yraw_{is}$ raw concentration of element$_i$ in source$_s$.

The adjusted R^2 of the multivariate regression equations are included in Eq. (5.3) as a weighting. This ensures that raw element concentrations have more weight in the correction of "less influenced" elements compared to "strongly influenced" elements.

The element data was divided into two data sets; surface sources S1–S3 with the three Wadi deposit samples (n = 21) and subsurface sources S4–S6 (n = 18). Since the latter consisted of crushed stones, a grain size analysis was not applied and only TOC and TIC were considered as influential parameters. However, Fig. 5.6 displays the mean TIC contents of sources S4–S6, which are located above the mean sink content. A data correction of TIC effects based on the regression Eqs. (5.1–5.3) would therefore imply an extrapolation of the data set. This extrapolation is associated with increasing inaccuracies in the prediction of value simulations and could therefore lead to overcorrections (Freund et al. 2006). We decided to determine correction factors for sources S4–S6 by using the Collins et al. (1997) method for TIC together with the determined R^2 values of the regression analysis in Eq. (5.2) for data correction.

5.3.5 *Element Selection and Mixing Model*

From the literature, we knew that all of the possible elements analysed have been identified in previous studies as potential fingerprinting properties and successfully implemented. Many rely on the conservativeness of P, ^{238}U, ^{226}Ra, ^{210}Pb and $^{210}Pb_{ex}$ (van Wijngaarden et al. 2002; Motha et al. 2002; Yeager et al. 2005; Collins and Walling 2007; Davis and Fox 2009; Evrard et al. 2011; Schuller et al. 2013). However, for Wadi Al-Arab, these elements were excluded prior to element selection because the geological unit Al-Hisa Phosphorites crops out in the Wadi bed (Fig. 5.1) and contains easily mobilised uranium and phosphorus (Moh'd and Powell 2010).

The procedure used in this study is based on commonly used selection tools in sediment fingerprinting (Collins et al. 2010; Davis and Fox 2009; Wilkinson et al. 2012), and was expanded by the inclusion of the water analysis results, a literature review and statistical analysis (grey shaded block; Fig. 5.4). This additional information ensured that the elements passing the mass conservation test and the Kruskal-Wallis test are surveyed a second time in regard to the corresponding element concentration in the water profile and characteristics mentioned in the

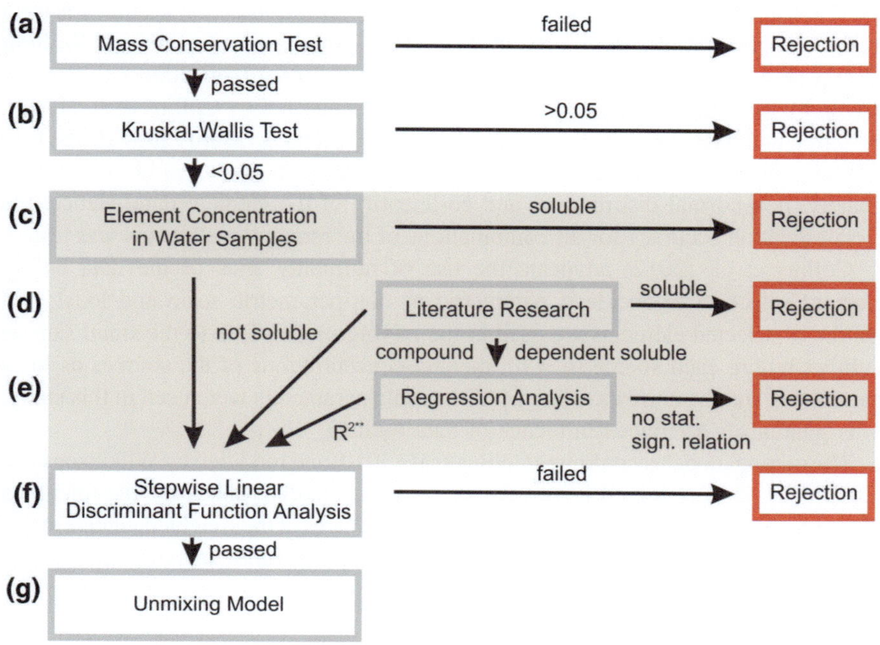

Fig. 5.4 The element selection approach

literature concerning their solubility under the given pH and Eh. If an element does not pass one of the tests, it is excluded from further procedures.

The mass conservation test (Fig. 5.4a) ensured that the element concentrations of the sink are located in between the sources with the highest and lowest median concentration plus/minus the corresponding median absolute deviation (MAD). A violation hints at the element's non-conservative behaviour and/or the absence of important sediment sources (Foster and Lees 2000; Haley 2010). The non-parametric Kruskal-Wallis (Fig. 5.4b) test identifies elements that can discriminate between at least two sources using the source median concentrations of an element (Collins and Walling 2002). The water analysis results serve as an additional indicator to eliminate the obvious soluble elements and reach a quick reduction of elements for the following literature review (Fig. 5.4c). Therefore, the element concentration of the sediments (mg kg^{-1}) and the element concentration of the water samples (mg l^{-1}) were converted to ppm. Following the simple assumption, if an element is soluble then the content (ppm) in the water will be enriched by >1 % compared to the sediment content (ppm). Elements that did not show any peculiarities or were not analysed in the water samples were evaluated in regard to the prevailing pH and Eh conditions, following the literature. All elements that were described as soluble were excluded (Fig. 5.4d). All elements that are compound-dependent soluble were correlated with the respective elements and were rejected in cases of a two-tailed significant relationship.

The remaining elements entered a stepwise linear discrimination function analysis (Fig. 5.4e) to find the best source discriminating composite fingerprint based on the raw data. The element combination with highest reclassification accuracy is used as the composite fingerprint in the mixing model proposed by Collins and Walling (2002) (Fig. 5.4e and Eq. 5.4). Instead of using a forward selection of elements driven by the minimisation of Wilks' lambda, as suggested by Collins et al. (1998), we agree with Haley (2010) that Wilks' lambda may not necessarily provide the best element combination in view of the reclassification result due to variance inhomogeneity, non-normal distribution, and co-linearity of the element data. Hence, the reclassification accuracy for all combinations of the remaining elements was tested.

Collins et al. (2012) advocate the use of normality tests of the data before element selection to decide if parametric or non-parametric scale and local estimators of selected elements are used in the mixing model. Due to the small sample number within each source (n = 6), median concentrations of the sources as local parameters and the corresponding MAD as scale parameters were used in this study. This guarantees a reduced influence of data outliers.

The mixing model according to Collins et al. (2010) and modified by Wilkinson et al. (2012); Eq. (5.4) is solved iteratively by inserting random values for the unknown source composition parts P_1 to P_m until the sum of squares of the weighted relative errors (ε) is minimized. This process is implemented under the following constrains:

$$P_t \geq 0$$
$$\sum P_t = 1$$
$$\varepsilon = \sum_{i=1}^{n} \left\{ \left(S_i - \left(\sum_{t=1}^{m} P_s V_{is} Yc_{is} \right) \right) / S_i \right\}^2 W_i \tag{5.4}$$

where:

ε sum of squares of the weighted relative errors;
n number of elements used in the composite fingerprint;
m number of sediment sources;
S_i sink concentration of element$_i$;
P_s unknown composition part of sediment source$_s$;
V_{is} within-source-variability weighting of element$_i$ and source$_s$;
Yc_{is} corrected concentration of element$_i$ in source$_s$; and
W_i discriminatory weighting of element$_i$.

To account for the heterogeneous nature of sediment sources, the model was not only solved for the mean element source concentrations, but also framed by a Monte Carlo sampling procedure giving information about the uncertainty of the source contributions based on the variability in element concentrations of the sources (Rowan et al. 2000; Small et al. 2002; Krause et al. 2003; Collins et al. 2010; Wilkinson 2012). Student's *t*-distributions were computed for each element in each source by using the source medians as local estimators and the corresponding MAD as the range of the corrected element data (Yc_{is}). The mean element concentrations of the core samples reflect the sink concentrations in the mixing model. The mixing model was then solved 2000 times, ensuring a homogenous distribution, with randomly generated element fractions of Yc_{is} within the two-tailed 95 % confidence interval of the Student's *t*-distributions. The mean of the 2000 mixing model results defines the source composition of the reservoir sediment and the standard deviation reflects the source uncertainties.

For providing information about the reliability of the estimated source composition, the mean relative error (MRE; Wilkinson et al. 2012) was used:

$$MRE = \frac{1}{r} \sum_{j=1}^{r} \left(\frac{\varepsilon}{n} \right) \tag{5.5}$$

where:

r number of iterations;
n number of sources; and
ε sum of squares of the relative element equation errors (Eq. 5.4).

The data correction and sediment fingerprinting was mainly applied using the open source statistic program R (R Development Core Team 2011). Table S1 in the Electronic Supplementary Material provides an overview of the R-packages and functions used.

5.4 Results

5.4.1 Sink Chemistry, Sediments and Influencing Parameters

The physiochemistry of the Wadi Al-Arab reservoir showed an abrupt decrease from an alkaline (pH 8.2) to a neutral pH (pH 7.29) as well as a drop of redox potential to <300 mV below 8 m water depth, indicating anaerobic conditions in the lower water body of the meromictic lake (Beutel et al. 2007). The reservoir sediment cores showed neither vertically nor spatially significant variability in the element and isotope concentrations. Thus, the mean tracer concentrations of all core sediment samples were used in the subsequent analysis.

The grain size analysis from the sources and the sink showed a clear trend in grain size selectivity (Fig. 5.5). Sand accumulates mostly in the Wadi deposits. On the other hand, clay and silt fractions are mainly transported through to the reservoir as suspended sediment yield. The TOC and TIC contents of the sources differ from the mean contents of the reservoir sediment (Fig. 5.6). Hence, grain size selectivity and substantial differences in organic matter and carbonate content are influencing parameters prohibiting the direct comparison between sink and source samples.

5.4.2 Data Correction for Sources 1 to 3 and Wadi Deposits

A total of 27 elements were analysed with SMRAM for the influence of the influential parameters (i.e. sand, silt, clay, TOC, TIC). Table S2 in the Electronic Supplementary Material summarises the results performed on the pedogenetic sources S1–S3 and the Wadi deposit samples. Significant relationships during the initial regression analysis are highlighted and show that clay (2), TOC (7) and TIC (14) each have an influence on 23 of 27 elements. Ni is not influenced by any of the tested parameters, whereas Ba, Br, Cr, Cu, Na, P, Sr, Zn, ^{210}Pb and ^{137}Cs are only

Fig. 5.5 Mean grain size contents between sources S1–S3, and Wadi deposits and sink

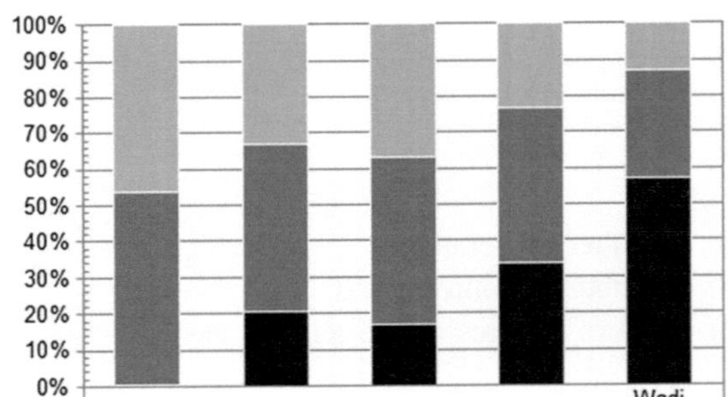

	Core 4 (n=8)	Source1 (n=6)	Source 2 (n=6)	Source 3 (n=6)	Wadi deposit (n=3)
Clay [%]	46.07	32.98	37.06	23.43	13.09
Silt [%]	53.63	46.27	46.17	42.89	30.06
Sand [%]	0.3	20.75	16.77	33.68	56.86

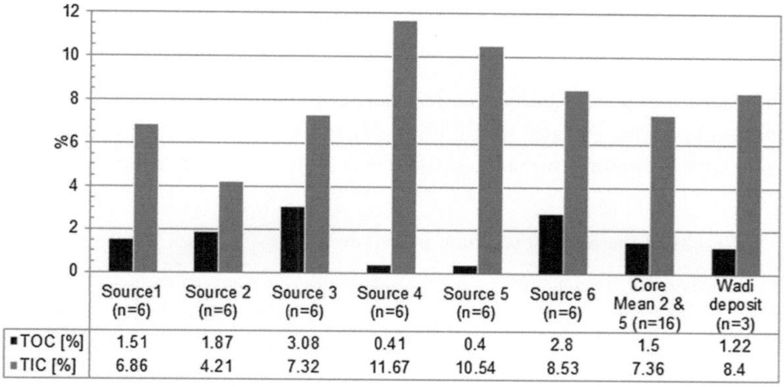

Fig. 5.6 Mean TOC and TIC contents for all sources, sink and the Wadi deposits

influenced by one of the aforementioned influential parameters. Furthermore, it is noticeable that TOC related elements are not influenced by sand, silt, clay and/or TIC, and vice versa. Apart from Ca and Sr, the direction of dependency between TIC and the related elements is negative, as is the dependency between sand and the related elements. Clay (with the exception of P) and TOC are always positively related with the elements. This study showed that the silt content of a sample has no significant influence on the element concentrations.

In the first-order semi-partial regression analysis, no further significant relationships were observed between the residuals from the initial regression analysis and the remaining influential parameters. Hence, the equations of the initial regression analysis serve as element correction functions (Eq. 5.1) and the corresponding R^2 values serve as goodness of fit for the corrected data set development (Eq. 5.3).

Table 5.5 shows the raw and corrected element concentrations following the SMARM in comparison to the method by Collins et al. (1997) using the SSA. Furthermore, we added the within-source variability weighting used in the mixing model. The SMARM corrected values have a higher similarity to the original concentrations and are therefore more environmentally sensitive compared to the SSA method.

5.4.3 *Data Correction for Sources S4 to S6*

No significant relationships could be identified between TOC and the elements using the SMRAM, while TIC is positive linear related with Ca ($R^2 = 0.97$) and negative linear related with Cr ($R^2 = 0.39$), Ni ($R^2 = 0.72$), P ($R^2 = 0.54$), Si ($R^2 = 0.82$), and Zn ($R^2 = 0.54$) at p-values <0.01.

5.4.4 Element Selection

Table 5.3 summarizes the selection procedure and shows that Zn failed the mass conservation test. Ba, Br, and Pb failed the Kruskal-Wallis test. The comparison of water and sink element content identified Na as non-conservative since the water

Table 5.3 Results of the element selection procedure

Element		Mass conservation test (Fig. 5.3a)			Kruskal-Wallis *H*-test (Fig. 5.3b)		Concentration in water (Fig. 5.3c)		Literature & correlation (Fig. 5.3d)	
		Source					Ø Water cont.[c] (mg/l)	Ø Water cont. (%)	Conservative elements enter linear discriminant analysis	
		Min[a]	Max[a]	Sink[b]	Chi2	*p*-value				
Al	(ppm)	2000	33,913	30,267	31	0.000	n.m.	/	cons.	Al
Ba		24	2268	573	**10**	**0.080****	/	/	/	Ba
Br		4	21	10	**9**	**0.093****	/	/	/	Br
Ca		162,497	321,371	246,406	27	0.000	74	0	**n.cons.**	Ca
Cr		11	238	223	21	0.001	n.m.	/	cons.	Cr
Cu		16	60	54	18	0.003	n.m.	/	cons.	Cu
Fe		350	25,256	20,002	31	0.000	0	0	cons.	Fe
K		140	5030	3690	27	0.000	10	0	**n.cons.**	K
Mg		1195	11,003	7007	24	0.000	38	1	**n.cons.**	Mg
Mn		8	832	316	28	0.000	0	0	cons.	Mn
Na		305	1156	1006	12	0.036	110	**10.93*****	/	Na
Ni		4	182	97	15	0.007	n.m.	/	cons.	Ni
Pb		2	12	11	**10**	**0.061****	/	/	/	Pb
Si		8391	152,066	88,680	27	0.000	9	0	cons.	Si
Sr		245	976	675	13	0.019	1	0	**n.cons.**	Sr
Ti		155	3258	2244	31	0.000	n.m.	/	cons.	Ti
Zn		**31**	**203**	**229***	/	/	/	/	/	Zn
Zr		5	152	80	31	0.000	n.m.	/	cons.	Zr
^{137}Cs		0	28	6	33	0.000	n.m.	/	cons.	^{137}Cs
^{40}K		1	144	114	26	0.000	n.m.	/	**n.cons.**	^{40}K
^{228}Th		1	18	15	31	0.000	n.m.	/	cons.	^{228}Th
^{232}Th		1	17	10	30	0.000	n.m.	/	cons.	^{232}Th
TOC (%)		0	4	2	24	0.000	n.m.	/	**n.cons.**	TOC

[a]Medians of corrected elements ± MAD
[b]Ø Concentration of reservoir core samples [n = 35]
[c]Mean content of water profiles [n = 17]
*Rejection if value is outside max and min range
**Rejection if *p*-value > 0.05
***Rejection if >1 %
/ Rejected by previous test
n.m. = Not measured
n.cons. = Not conservative
cons. = Conservative

samples were enriched by 10.93 % compared to the sediment concentration. The concentrations of Ca, Fe, K, Mg, Mn, Si, and Sr were <1 % and together with the radionuclides they entered the literature review stage (d in Fig. 5.4) concerning pH and Eh. Horowitz (1991) and Scheffer and Schachtschabel (2010) state that for an alkaline to neutral pH and a relatively high mean content of TIC—as found in the reservoir sediments—Al, Cr, Cu and Ni behave conservatively. Fe and Mn-oxihydroxides are, besides clay, important absorption partners for cations (Horowitz 1991; Slattery et al. 1999), which may become recycled under anaerobic conditions. However, in the Al-Arab reservoir, the pH never falls below 7.29 and Eh always exceeds +127 mV (unpublished data, Saadoun et al. 2010), resulting in stable conditions (Scheffer and Schachtschabel 2010) and, hence, Mn and Fe can be considered as conservative. The oxides and associated minerals of Ti, Zr and Th (^{232}Th, ^{228}Th) are highly insoluble (Adriano 1986; Scheffer and Schachtschabel 2010) and consequently conservative, as is ^{137}Cs (Motha et al. 2002).

Due to high phytoplankton production during winter (Saadoun et al. 2008), and subsequent in-situ sedimentation of organic matter, TOC is excluded from the element selection. Furthermore, Adriano (1986), Horowitz (1991) and Scheffer and Schachtschabel (2010) reported that the nutrients K (including ^{40}K) and Mg are possibly subject to leaching. Thus, they were excluded from the selection procedure as a precautionary measure. Overall, 12 components (Al, Cr, Cu, Fe, Mn, Ni, Si, Ti, Zr, ^{137}Cs, ^{232}Th, and ^{228}Th) proved to be conservative and were used for the stepwise linear discriminant analysis.

The combination Al, Cr, Fe, Mn, ^{137}Cs, ^{232}Th and ^{228}Th provided the best source discrimination with a cross-validated accuracy rate of 94.4 %. One sample of source S6 and one sample of source S5 were misclassified to source S4. All surface sources (S1–S3) were correctly classified during cross-validation (Tables 5.4, 5.5).

Table 5.4 Source prediction of the linear discriminant analysis with Al, Cr, Mn, Fe, ^{232}Th, ^{228}Th, ^{137}Cs

	Source prediction with Al, Cr, Mn, Fe, ^{232}Th, ^{228}Th, ^{137}Cs						Element	%[a]	TDW[b]
Source	S1	S2	S3	S4	S5	S6	Al	58.33	1.5
S1	6 (100 %)	0	0	0	0	0	Cr	50	1.3
S2	0	6 (100 %)	0	0	0	0	Mn	38.89	1
S3	0	0	6 (100 %)	0	0	0	Fe	63.89	1.6
S4	0	0	0	6 (100 %)	0	0	^{232}Th	58.33	1.5
S5	0	0	0	1 (16.7 %)	5 (83.3 %)	0	^{228}Th	61.11	1.6
S6	0	0	0	1 (16.7 %)	0	5 (83.3 %)	^{137}Cs	58.33	1.5

[a]Source samples predicted correctly by individual elements

[b]Element discriminatory weighting used in unmixing model

Table 5.5 Comparison of uncorrected, SMARM-corrected and SSA-corrected values

Source	Al (ppm)					Cr (ppm)				
	Raw median	Corr. median (SMARM)	Corr. median (Collins et al. 1997)	CV %[b]	SV^{c}_{si}	Raw median	Corr. median (SMARM)	Corr. median (Collins et al. 1997)	CV %[b]	SV^{c}_{si}
S1	25045	22114	32395	18.11	0.55	104	104	135	28.85	0.36
S2	51645	28537	26699	18.84	0.53	169	158	87	12.17	0.85
S3	27360	27070	24148	9.92	1.00	219	185	193	28.38	0.36
S4	2285	2285	13759	12.47	0.80	17	20	102	41.18	0.25
S5	9260	9260	60132	20.19	0.49	117	130	757	48.07	0.21
S6	3970	3970	20149	30.48	0.33	136	142	690	10.29	1.00
Core[a]	30267					223				
Source	Mn (ppm)					Fe (ppm)				
	Raw median	Corr. median (SMARM)	Corr. median (Collins et al. 1997)	CV %[b]	SV^{c}_{si}	Raw median	Corr. median (SMARM)	Corr. median (Collins et al. 1997)	CV %[b]	SV^{c}_{si}
S1	623	589	806	41.33	0.49	17660	15421	22843	16.02	0.92
S2	743	575	384	23.91	0.85	40325	21396	20847	18.04	0.82
S3	251	250	222	20.32	1.00	17490	17288	15437	14.78	1.00
S4	21	21	123	21.95	0.93	770	770	4636	54.55	0.27
S5	33	33	214	22.73	0.89	6960	6960	45196	18.53	0.80
S6	12	12	58	30.43	0.67	1610	1610	8171	45.65	0.32
Core[a]	316					20,002				
Source	^{232}Th (Bq kg^{-1})					^{228}Th (Bq kg^{-1})				
	Raw median	Corr. median (SMARM)	Corr. median (Collins et al. 1997)	CV %[b]	SV^{c}_{si}	Raw median	Corr. median (SMARM)	Corr. median (Collins et al. 1997)	CV %[b]	SV^{c}_{si}

(continued)

Table 5.5 (continued)

Source	^{232}Th (Bq kg^{-1})					^{228}Th (Bq kg^{-1})				
	Raw median	Corr. median (SMARM)	Corr. median (Collins et al. 1997)	CV %[b]	SV^{c}_{si}	Raw median	Corr. median (SMARM)	Corr. median (Collins et al. 1997)	CV %[b]	SV^{c}_{si}
S1	14.23	12.42	18.40	8.11	1.00	13.96	12.42	18.10	14.51	0.90
S2	26.11	14.26	13.50	21.14	0.38	25.83	14.90	13.40	20.99	0.62
S3	11.80	11.65	10.40	15.85	0.51	12.12	12.00	10.70	13.08	1.00
S4	1.13	1.10	6.80	11.87	0.68	1.26	1.00	7.60	32.00	0.41
S5	3.74	4.04	24.30	26.89	0.30	4.16	3.49	27.00	37.02	0.35
S6	2.74	1.81	13.90	22.44	0.36	1.84	1.80	9.30	32.12	0.41
Core[a]	9.59					15.08				

Source	^{137}Cs (Bq kg^{-1})				
	Raw median	Corr. median (SMARM)	Corr. median (Collins et al. 1997)	CV %[b]	SV^{c}_{si}
S1	17.9	17.9	23.2	30.18	0.31
S2	9.39	8.43	4.90	9.43	1.00
S3	30.3	22.5	26.7	23.68	0.40
S4	0.00	0.00	0.00	0.00	1.00
S5	0.00	0.00	0.00	0.00	1.00
S6	0.00	0.00	0.00	0.00	1.00
Core[a]	5.73				

The SMARM-corrected values were used for the Monte Carlo sampling and mixing model

[a]Mean raw tracer concentrations of 38 core sediment samples

[b]Derivated from the median absolute deviations of the source samples

[c]Within-source-variability weighting for the respective element and source used in the mixing model, determined from the median absolute deviations

5.4.5 *Mixing Model and Monte Carlo Uncertainty Estimation*

The results of the mixing model indicate that 59 ± 8 % of the sampled reservoir sediments are delivered from sloped cultivated olive orchards (S1). Fields for agriculture on plateau and saddle positions (S2) are associated with 11 ± 9 %, with 29 ± 15 % derived from slopes used for grazing (S3), and 1 ± 1.4 % derived from the geological unit Al-Hisa Phosphorites/Amman Silicified Limestone (S6). However, with reference to the error, no sediment originated from the geological sources (S4–S6). It was possible to reconstruct the mean element concentrations of the sink with a mean residual error of 1.04 %.

5.5 Discussion

5.5.1 *Data Correction and Element Selection*

Results showed that the grain size composition of the <63 μm fraction of source and sink sediments was substantially different (Fig. 5.5). Although Walling (2013) maintains that most sediments transported are <63 μm, the change in fraction volume would have prohibited a direct comparison of source and sink sediments, rendering the additional implementation of correction factors necessary. Sampling smaller grain sizes (<10 μm) as suggested by Wilkinson et al. (2012) implies that the source affiliation is only limited to this fraction and demands proof that the chosen fraction represents the major component of the sink sediment (Walling 2013). This was not the case for the Al-Arab reservoir sediments since the major component was silt (<63 μm).

Koiter et al. (2013a, b) shows that in catchments where disintegration and/or attrition of sediments during transport is likely, the general assumption that finer-grained material is more chemically reactive can be an incorrect conclusion and thus selective sampling of smaller grain size fractions is inappropriate. This might be of special relevance to carbonatic catchments, such as the Wadi Al-Arab, where the disintegration of carbonatic sand particles needs to be considered. However, we recognize that in certain cases, and with the necessary prior knowledge about the composition of the grain size fractions before and during transport as well as after deposition, the selective sampling approach has the potential to give better accuracy to the results than any numerical correction. Nevertheless, as in most cases, the work implemented in north Jordan was a pilot study and no prior knowledge existed as to grain size fractions, particle selectivity and organic or inorganic carbon content. Selective grain size sampling or the removal of organic material would have implied the risk of neglecting important information beforehand and, with regards to the results, would have followed incorrect assumptions.

Table 5.6 Regression coefficient of influencing parameters (influential parameters; Spearman correlation)

n = 21[a]	TOC	TIC	Clay	Silt
TIC	−0.21			
Clay	−0.15	−0.73**		
Silt	−0.07	−0.25	0.23	
Sand	0.33	0.67**	−0.91**	−0.46*

[a]18 samples of source S1–S3 and 3 samples of the Wadi deposits
*significant with p-value < 0.05
**significant with p-value < 0.01

The SMRAM confirms that sand and TIC generally dilute (negatively correlate) a sample, whereas TOC and clay enrich the concentration, corresponding to known literature (Adriano 1986; Horowitz 1991; Scheffer and Schachtschabel 2010). The positive relation of TIC with Ca ($R^2 = 0.99$) and Sr ($R^2 = 0.42$) leads to the suggestion that the TIC content, and thus the bedrock in the catchment, mainly consists of calcium carbonate and partly of strontium carbonate. The method further revealed that TIC, from all tested influential parameters, showed the highest R^2 values and best explains the variance of the 14 out of 27 related elements. Thus, the inclusion of TIC as an influencing parameter is judged as essential for carbonatic areas, as was recommended by Horowitz (1991). More importantly, the SMRAM revealed the absence of any further relationships after the first regression analysis, explained by the interrelations of the influential parameters (Table 5.6). Clay and TIC show a significant negative relationship, while sand and TIC have a significant positive one. This indicates that the sand fraction is made of carbonatic particles. TOC is free of interrelations with any other influential parameter. Collins et al. (1997), Russell et al. (2001) and Mukundan (2012) report a general influence by TOC, which is most probably closely related to grain size influence, so one has to be careful of over correction when addressing both.

In north Jordan, existing uranium and phosphorus deposits in the bedrock prevent the use of P, ^{238}U, ^{226}Ra, ^{210}Pb, and $^{210}Pb_{ex}$ as fingerprinting elements, because they can enter the system in additional, undetermined amounts. Additionally, we tested the elements using the element selection procedure for reassurance. Results affirmed the decision for exclusion. ^{238}U, ^{226}Ra, ^{210}Pb failed the mass conservation test, due to unknown enrichment. The $^{210}Pb_{ex}$ core values were mainly negative because the unknown accumulation in ^{226}Ra disturbs the comparability with ^{210}Pb; both form the basis of the atmospheric $^{210}Pb_{ex}$ calculation. Phosphorus was excluded because of its possible solubility (Scheffer and Schachtschabel 2010). The results generally suggest that a profound knowledge of the geological background of the study area is of advantage for the choice of suitable fingerprinting elements.

The water analysis results provided important information on the pH and Eh conditions in the water profile, thus allowing evaluation of the changing environmental conditions that the sediments go through while settling. Including the water element concentrations, the literature review and the correlation analysis into the expanded selection procedure helped to quickly eliminate elements whose conservative behaviour in water could be doubted. Results showed that the mass conservation and Kruskal-Wallis test did not suffice in rejecting the elements Ca, Na and Sr, which are undoubtedly soluble in water, but have been used as conservative fingerprinting properties in other studies analyzing suspended sediment sources (e.g. Russell et al. 2001; Collins and Walling 2002; Carter et al. 2003; de Miguel et al. 2005). Furthermore, the problem of phytoplankton production in lakes is quite common and prohibits the use of organic properties, such as TOC. Furthermore, K and Mg are potentially soluble and were therefore removed by the expanded procedure, although they are, together with TOC, considered as conservative fingerprinting properties in other studies using lake sediments (Kelly and Nater 2000; Kouhpeima et al. 2010; Slimane et al. 2013).

5.5.2 Results of the Sediment Fingerprinting

Despite the uncertainties associated with the source composition, the results clearly show that steep slopes with shallow, yellow, Mediterranean soils underlying cultivated olive orchards (S1 = 59 $\pm$ 8 %) and grazing land (S3 = 29 $\pm$ 15 %, Table 5.1), when combined, contributed $\sim$90 $\pm$ 17 % of the reservoir sediments. The marly substrate of the yellow Mediterranean soils have a high Na content, leading to unstable aggregates and surface crusting, concurrently delivering good conditions for runoff generation, favoured by the fragmented vegetation cover (Yassoglou and Kosmas 1997; Thornes 2005; García-Ruiz 2010; unpublished field observations). These developments are of special relevance within the catchment on the south facing slopes as soil temperature and rain shadow visibly enhance this development strongly. These unfavourable characteristics of the present soils, together with the steep slope gradient and agricultural use, trigger the high soil erosion in these land units.

Our erosion measurements in seven olive orchards on slopes >10 % showed very high potential erosion rates (95 $\pm$ 8 t ha^{-1} yr^{-1}) mainly attributed to the downhill ploughing, the canalisation of runoff water and, consequently, the triggering of excessive water erosion (Fig. 5.2a; Kraushaar et al. 2014). These findings are supported by other studies from the Mediterranean on conventionally tilled olive orchards using different methods of investigation (Gómez et al. 2003; Milgroom et al. 2005; Bruggemann et al. 2005; Francia Martínez et al. 2006; Ramos et al. 2008; Barnevald et al. 2009; Vanwalleghem et al. 2010, 2011). Slopes that are not ploughed but undergo grazing are likely to suffer from different processes including, thinning of vegetative coverage, compaction of soil surface, reduced infiltration, and therefore increased runoff (Yassoglou and Kosmas 1997). Furthermore, as with Wadi Al-Arab, substantial evidence of mechanical translocation of the stony mulch layer that protects the surface is highly evident. A typical pattern of sheep and goat tracks exists on the majority of the slopes, leaving a reduced resistance to water erosion (Fig. 5.2c).

Fields used for agriculture on plateau and saddle positions that contain red Mediterranean soils (S2) show a much lower contribution to the sink at only

11 ± 9 %. The existence of red Mediterranean soils with their various pedogentic features (Yaalon 1997) and their advanced soil development can be interpreted as proof of stable geomorphic conditions (Verheye and de la Rosa 2005; Federoff and Courty 2013). However, recent changes in management caused shallow gullies in the red soils throughout the catchment. Broken road boundaries or the clearance of vegetation strips channelize and accelerate surface water runoff in upstream positions, increase transportation capacity, and cause gullying, even in almost level locations. Typically, soils in grain fields are unprotected during the rainy season (October–February) and are, therefore, highly vulnerable to extreme rainfall events causing rills and shallow ephermal gully erosion (García-Ruiz 2010). These results are supported by Khresat and Taimeh (1998) who reports water erosion and improper farming practices to be the major causes of physical soil degradation in the study area.

Sources 4–6 were, with regard to the error, not detected in the sink as these sediments do not contain ^{137}Cs that was present in the lake sediments. These sources were chosen to represent subsurface samples as C horizons from gully erosion, in addition to new sediments derived from construction work on roads where enormous masses of unconsolidated bedrock sediments are produced and dumped onto the slopes. However, with regards to the fingerprinting data, either they did not deliver enough material to the sink for the model to detect, and/or the bedrock samples do not represent these sources well enough. Future research should focus on sampling the aforementioned sources directly, and consider the vertical sampling of profiles to differentiate topsoil from gully erosion for their relative contribution to the sink.

5.5.3 *Implications for the Wadi Al-Arab Reservoir*

The reservoir experienced ~9 % loss of total water storage volume in 26 years. The clayey and silty sediments accumulate at the wall of the dam near the water outlet. Here, the sediment line is around 1–2 m below the outlet, posing an acute threat to the reservoir as there is no available removal equipment (Hussein Al-Shurieki 2011, personal communication, dam director).

Particle selectivity leads to an enrichment of the heavy metals, such as chromium (228 mg kg^{-1}), nickel (99.5 mg kg^{-1}) and cadmium (4 mg kg^{-1}), in the reservoir sediments, exceeding the probable effect level for an ecosystem (USEPA 2014: Cr = 90 mg kg^{-1}; Ni = 36 mg kg^{-1}; Cd = 3.53 mg kg^{-1}). Due to the amount of clay, the high concentration of iron, and carbonatic background (mean TIC = 7.4 %), and thus a pH > 7, the heavy metals are considered as immobile in the reservoir sediments (Blume 2004). However, sediments should not be hauled to the surface and used as agricultural ammendments since heavy metals might become mobile under changing conditions, such as agricultural use. Thus, the only effective way of managing the reservoir is to prevent sediments from entering it, with counter erosion measurements on the slopes, predominantly in the most affected source areas, such as olive orchards and slopes used for grazing.

5.6 Conclusions

A sediment fingerprinting procedure was implemented for a Mediterranean to semi-arid catchment in north Jordan with six potential sediment sources. On the basis of this pilot study, an adapted correction factor approach is suggested to guarantee comparability of sink and source samples. This approach includes the exploration of possible linear, as well as non-linear, positive and negative relationships between individual elements, radionuclides and the identified influential parameters of clay, silt, sand, TOC and TIC. Correction factors were developed that incorporate the concept of interrelations in the calculation of element-specific correction factors. Results highlight the influential impacts of TIC, particularly in the carbonatic Wadi Al-Arab catchment, and the majority of the elements needed to be corrected for differences in TIC content. TOC did not show interrelations to other influential parameters, such as clay, silt or sand. However, when more than one influential parameter existed (usually clay, sand and TIC), the correction of TIC was sufficient to erase the influence of the other parameters due to interrelations between them. Finally, Ni did not need any correction. The strong influence of TIC content is assumed to exist in regions of carbonatic bedrock as well as in areas of carbonatic dust input. Carbonate-free areas should be free on any TIC influence on the element concentration.

The correction factors are calculated element—but not source-specific since the regression analyses are sensitive to outliers in small data sets. The general trend (positive or negative) between the influential parameters and the element are assumed to be similar for each source. Nevertheless, it would be of great advantage to further explore source-specific relations with an expanded data set.

The element selection procedure should not rely solely on a cross check of the element's application in other, similar research studies, nor the conventional element selection procedure comprising a mass conservation test and the Kruskal-Wallis test. Both approaches are not enough to reliably eliminate non-conservative elements, especially when considering reservoir or lake sediments. Additional physiochemical information on the sink environment—a literature review in this case—and correlation analysis helped to reject further elements. However, the study also showed that an additional comprehensive and conservative literature review with consideration of prevailing pH and Eh conditions alone would have eliminated these respective elements. When using lake or reservoir sediments in sediment fingerprinting, organic elements need to be carefully evaluated for their conservativeness due to possible phytoplankton production, and thus an unspecified additional organic source.

We conclude from the analysis of the lake sediments that the element concentrations did not vary significantly in the core samples and throughout the lake, and thus indicate similar source contributions throughout the 26 years of the reservoir's existence. Furthermore, no exhaustion of the identified surface sources, and the related human impact on the system, could be detected. The observed sediment delivery to the reservoir, and associated sedimentation, has resulted in an acute threat to the reservoir's life span.

Acknowledgements The authors would like to thank A. Rämmler, J. Fröhlich, D. Sonntag, D. Bednorz, I. Volkmann, H-J. Stärk, C. Bönisch, M. Wunderlich and S. Mothes (UFZ), M. Weber (MLU Halle) and B. Winkler (TU Dresden) for analytics and interpretation advice. We are grateful to H. Al-Shurieki (Al-Arab reservoir) and M. Raggad and E. Salameh (Jordan Univ.) for their steady support on site. We also wish to thank the German tax payer for funding the SMART II project (FKZ 02-WM1080) through the Federal Ministry of Education and Research, the IPSWat Program, and Helmholtz Interdisciplinary Graduate School for Environmental Research (HIGRADE).

References

Adriano DC (1986) Trace elements in the terrestrial environment. Springer, New York

Al-Ansari N, Shatnawi A (2011) Siltation of three small reservoirs in Jordan. Publications by the Institute of Earth and Environmental Sciences, Al Al-Bayt University, Jordan

Al-Ansari N, Alroubai A, Knutsson S (2012) Bathymetry and sediment survey for two old water harvesting schemes, Jordan. J Earth Sci Geotech Engineer 2:13–23

Al-Sharhan AS, Nairn AEM (1997) Sedimentary basins and petroleum geology of the middle East. Elsevier, Amsterdam

Barnevald RJ, Bruggeman A, Sterk G, Turkelboom F (2009) Comparison of two methods for quantification of tillage erosion rates in olive orchards of north-west Syria. Soil Till Res 103:105–112

Beutel MW, Leonarda TM, Denta SR, Mooreb BC (2007) Effects of aerobic and anaerobic conditions on P, N, Fe, Mn, and Hg accumulation in waters overlaying profundal sediments of an oligo-mesotrophic lake. Water Res 42:1953–1962

Blume HP (ed) (2004) Handbuch des Bodenschutzes. Bodenökologie und –belastung - Vorbeugende und abwehrende Schutzmaßnahmen, 3rd edition. ecomed, Landsberg am Lech, pp 723–769

Bruggemann A, Masri Z, Turkelboom F (2005) Strategies to sustain productivity of olive groves on steep slopes in the northwest of the Syrian Arab Republic. In: Benites J, Pisante M, Stagnari F (eds) Integrated soil and water management for orchard development; role and importance. FAO Land and Water Bulletin 10. FAO, Rome, pp 75–87

Carter J, Owens PN, Walling DE, Graham JLL (2003) Fingerprinting suspended sediment sources in a large urban river system. Sci Total Environ 314–316:513–534

Collins AL, Walling DE (2002) Selecting fingerprint properties for discriminating potential suspended sediment sources in river basins. J Hydrol 261:218–244

Collins AL, Walling DE (2007) Sources of fine sediment recovered from the channel bed of lowland groundwater-fed catchments in the UK. Geomorphology 88:120–138

Collins AL, Walling DE, Leeks GJL (1997) Source type ascription for fluvial suspended sediment based on a quantitative composite fingerprinting technique. Catena 29:1–27

Collins AL, Walling DE, Leeks GJL (1998) Use of composite fingerprints to determine the provenance of the contemporary suspend sediment load transported by rivers. Earth Surf Process Landforms 23:31–52

Collins AL, Walling DE, Webb L, King P (2010) Apportioning catchment scale sediment sources using a modified composite fingerprinting technique incorporating property weightings and prior information. Geoderma 155:249–261

Collins AL, Zhang Y, McChesney D, Walling DE, Haley SM, Smith P (2012) Sediment source tracing in a lowland agricultural catchment in southern England using a modified procedure combining statistical analysis and numerical modelling. Sci Total Environ 414:301–317

Davis CM, Fox JF (2009) Sediment fingerprinting: review of the method and future improvements for allocating nonpoint source pollution. J Environ Eng 135:490–504

De Miguel E, Charlesworth S, Ordonez A, Seijas E (2005) Geochemical fingerprints and controls in the sediments of an urban river: river Manzanares, Madrid (Spain). Sci Total Environ 340:137–148

Döpke G (2004) Regionale und substratabhängige Verteilung von Schwermetallen in oberflächennahen Sedimenten der Inner Kingston Basin, Ontariosee. Dissertation, University of Osnabrück, Austria

El-Radaideh NM (2010) Using bottom reservoir sediments as a source of agricultural soil: Wadi El-Arab Reservoir as a case study, NW Jordan. ABHATH AL-YARMOUK Basic Sci Eng 19:75–91

Evrard O, Navratil O, Ayrault S, Ahmadi M, Némery J, Legout C, Lefèvre I, Poirel A, Bonté P, Esteves M (2011) Combining suspended sediment monitoring and fingerprinting to determine the spatial origin of fine sediment in a mountainous river catchment. Earth Surf Process Landforms 36:1072–1089

Federoff N, Courty M-A (2013) Revisiting the genesis of red Mediterranean soils. Turkish J Earth Sci 22:359–375

Foster IDL, Lees JA (2000) Tracers in geomorphology—theory and applications in tracing fine particulate sediments. In: Foster IDL (ed) Tracers in geomorphology. Wiley, Chichester, pp 3–21

Foster IDL, Boardman J, Keay-Bright J (2007) Sediment tracing and environmental history for two small catchments, Karoo Uplands, South Africa. Geomorphology 90:126–143

Francia Martínez JR, Durán Zuazo VH, Martínez Raya A (2006) Environmental impact from mountainous olive orchards under different soil-management systems (SE Spain). Sci Total Environ 358:46–60

Freund RJ, Wilson WJ, Sa P (2006) Regression analysis—statistical modeling of a response variable, 2nd edn. Elsevier, Oxford

García-Ruiz JM (2010) The effects of land uses on soil erosion in Spain: A review. Catena 81:1–11

Ghrefat H, Yusuf N (2006) Assessing Mn, Fe, Cu, Zn, and Cd pollution in bottom sediments of Wadi Al-Arab Dam, Jordan. Chemosphere 65:2114–2121

Gómez JA, Battany B, Renschler CS, Fereres E (2003) Evaluating the impact of soil management on soil loss in olive orchards. Soil Use Manage 19:127–134

Haley S (2010) The application of sediment source fingerprinting techniques to river floodplain cores, to examine recent changes in sediment sources in selected UK river basins. Unpublished PhD thesis, University of Exeter, Exeter, UK

Horowitz AJ (1991) A primer on sediment-trace element chemistry, 2nd edn. Lewis Publishers, Chelsea

Kelley DW, Nater EA (2000) Source apportionment of lake bed sediments to watersheds in an Upper Mississippi basin using a chemical mass balance method. Catena 41:277–292

Khresat SA, Taimeh AY (1998) Properties and characterization of Vertisols developed on limestone in a semi-arid environment. J Arid Environ 40:235–244

Koiter AJ, Owens PN, Petticrew EL, Lobb DA (2013a) The behavioural characteristics of sediment properties and their implications for sediment fingerprinting as an approach for identifying sediment sources in river basins. Earth-Sci Rev 125:24–42

Koiter AJ, Lobb DA, Owens PN, Petticrew EL, Tiessen KHD, Li S (2013b) Investigating the role of connectivity and scale in assessing the sources of sediment in an agricultural watershed in the Canadian prairies using sediment source fingerprinting. J Soils Sed 13:1676–1691

Kouhpeima A, Hashemi SAA, Feiznia S, Ahmadi H (2010) Using sediment deposited in small reservoirs to quantify sediment yield in two small catchments of Iran. J Sustain Devel 3 (3):133–139

Krause AK, Franks SW, Kalma JD, Loughran RJ, Rowan JS (2003) Multi-parameter fingerprinting of sediment deposition in a small gullied catchment in SE Australia. Catena 53:327–348

Kraushaar S, Herrmann N, Ollesch G, Vogel HJ, Siebert C (2014) Mound measurements—quantifying medium term soil erosion under olive trees in Northern Jordan. Geomorphology 213:1–12

Kraushaar S, Schumann T, Ollesch G, Schubert M, Vogel H-J, Siebert C (2015) Sediment fingerprinting in northern Jordan: element-specific correctionfactors in a carbonatic setting. Journal of Soils and Sediments 15:2155–2173

Lucke B (2007) Demise of the decapolis. Past and present desertification in the context of soil development, land use and climate. Unpublished PhD thesis, Brandenburg Technical University of Cottbus, Germany

Milgroom J, Auxiliadora Soriano M, Garrido JM, Gómez JA, Fereres E (2005) The influence of a shift from conventional to organic olive farming on soil management and erosion risk in southern Spain. Renew Agr Food Sys 22:1–10

Miller JR, Lord M, Yurkovich S, Mackin G, Kolenbrander L (2005) Historical trends in sedimentation rates and sediment provenance, Fairfield Lake, eastern north Carolina. J Am Water Ressour Assoc 41(5):1053–1075

MoA (Ministry of Agriculture) (1994) National soil map and land use project—the soils of Jordan, vol 3. Hunting Technical Services LTD, Jordan (Level 2), Amman, Jordan

Moh'd BK (2000) The geology of Irbid and Ash Shuna Ash Shamaliyya (WAQQAS), Map sheets no. 3154-II and 3154-III. Geological Directorate. Geological Mapping Division, Bulletin 46, Amman, Jordan

Moh'd BK, Powell JH (2010) Uranium distribution in the upper Cretaceous-Tertiary Belqa Group, Yarmouk Valley, northwest Jordan. Jordan J Earth Environ Sci (JJEES) 3:49–62

Motha JA, Wallbrink PJ, Hairsine PB, Grayson RB (2002) Tracer properties of eroded sediment and source material. Hydrol Process 16:1983–2000

Motha JA, Wallbrink PJ, Hairsine PB, Grayson RB (2004) Unsealed roads as suspended sediment sources in an agricultural catchment in south-eastern Australia. J Hydrol 286:1–18

Mukundan R, Walling DE, Gellis AC, Slattery MC, Radcliffe DE (2012) Sediment source fingerprinting: transforming from a research tool to a management tool. J Am Water Resour Assoc (JAWRA) 48:1241–1257

Nortcliff S, Carr G, Potter RB, Darmame K (2008) Jordan´s water resources: challenges for the future. Geographical Paper No. 185, The University of Reading, Reading, UK

Owens PN, Xu Z (2011) Recent advances and future directions in soils and sediments. J Soils Sed 11:875–888

Palmieri A, Shah F, Dinar A (2001) Economics of reservoir sedimentation and sustainable management of dams. J Environ Manage 61:149–163

Ramos MI, Feito FR, Gil AJ, Cubillas JJ (2008) A study of spatial variability of soil loss with high resolution DEMs: a case study of a sloping olive grove in southern Spain. Geoderma 148:1–12

Rowan JS, Goodwill P, Franks SW (2000) Uncertainty estimation in fingerprinting suspended sediment sources. In: Foster IDL (ed) Tracers in geomorphology. Wiley, Chichester, pp 279–290

Russell MA, Walling DE, Hodkinson RA (2001) Suspended sediment sources in two small agricultural lowland catchments in the UK. J Hydrol 252:1–24

Saadoun I, Bataineh E, Al-Handal AY (2008) The primary production conditions of Wadi Al-Arab Dam (Reservoir), Jordan. Jordan J Biol Sci 1:67–72

Saadoun I, Batayneh E, Alhandal A, Hindieh M (2010) Physicochemical features of the Wadi Al-Arab Dam (reservoir), Jordan. Intern J Ocean Hydrobiol 39:191–205

Salomons W, Mook WG (1987) Natural tracers for sediment transport studies. Cont Shelf Res 7:1333–1343

Scheffer F, Schachtschabel P (2010) Lehrbuch der Bodenkunde, 16th edn. Spektrum Akademischer Verlag, Heidelberg
Schuller P, Walling DE, Iroumée A, Quilodránd C, Castilloa A, Navase A (2013) Using ^{137}Cs and ^{210}Pbex and other sediment source fingerprints to document suspended sediment sources in small forested catchments in south-central Chile. J Environ Radioact 124:147–159
Slattery WJ, Conyers MK, Aitken RL (1999) Soil pH, aluminium, manganese and lime requirement. In: Peverill KI, Sparrow LA, Reuter DJ (eds) Soil analysis—an interpretation manual. CSIRO, Collingwood, Australia, pp 103–128
Slimane AB, Raclot D, Evrard O, Sanaa M, Lefèvre I, Ahmadi M, Tounsi M, Rumpel C, Mammou AB, Le Bissonnais Y (2013) Fingerprinting sediment sources in the outlet reservoir of a hilly cultivated catchment in Tunisia. J Soils Sed 13:801–815
Small IF, Rowan JS, Franks SW (2002) Quantitative sediment fingerprinting using a Bayesian uncertainty estimation framework. In: Dyer FJ, Thoms MC, Olley JM (eds) The structure, function and management implications of fluvial sedimentary systems. IAHS Publ 276, IAHS Press, Wallingford, UK, pp 443–450
Smith HG, Blake WH (2014) Sediment fingerprinting in agricultural catchments: a critical re-examination of source discrimination and data corrections. Geomorphology 204:177–191
Stumm W, Morgan JJ (1996) Aquatic chemistry: chemical equilibria and rates in natural waters, 3rd edn. Wiley, New York
R Development Core Team (2011) R: a language and environment for statistical computing. R Foundation for Statistical Computing, Vienna, Austria. Available at: http://www.R-project.org. Accessed 13 Oct 2013
Thornes JB (2005) Coupling erosion, vegetation and grazing. Land Degrad Dev 16:127–138
United States Environment Protection Agency—USEPA (2014) Water quality standards handbook —chapter 3: water quality criteria (40 CFR 131.11) http://water.epa.gov/scitech/swguidance/standards/handbook/chapter03.cfm. Last checked 12.06.2014
Vanmaercke M, Poesen J, Verstraeten G, de Vente J, Ocakoglu F (2011) Sediment yield in Europe: spatial patterns and scale dependency. Geomorphology 130:142–161
Vanwalleghem T, Laguna A, Giráldez JV, Jiménez-Hornero FJ (2010) Applying a simple methodology to assess historical soil erosion in olive orchards. Geomorphology 114:294–302
Vanwalleghem T, InfanteAmte J, González deMolina M, Soto Fernández D, Gómez JA (2011) Quantifying the effect of historical soil management on soil erosion rates in Mediterranean olive orchards. Agric Ecosyst Environ 142:341–351
van Wijngaarden M, Venema LB, De Meijer RJ, Zwolsman JJG, Van Os B, Gieske JMJ (2002) Radiometric sand–mud characterisation in the Rhine-Meuse estuary Part A. Fingerprinting. Geomorphology 43:87–101
Verheye W, de la Rosa D (2005) Mediterranean soils. In: Encyclopedia of life support systems (EOLSS) developed under the Auspices of the UNESCO (ed) Land use and land cover. Eolss Publishers, Oxford, UK. http://digital.csic.es/bitstream/10261/38282/1/Mediterranean%20soils.pdf. Accessed 3 May 2015
Walling DE (2005) Tracing suspended sediment sources in catchments and river systems. Sci Total Environ 344:159–184
Walling DE (2013) The evolution of sediment source fingerprinting investigations in fluvial systems. J Soils Sed 13:1658–1675
Walling DE, Woodward JC, Nicholas AP (1993) A multiparameter approach to fingerprinting suspended sediment sources. In: Peters NE, Hoehn E, Liebundgut C, Tase N, Walling DE (eds) Tracers in hydrology. IAHS Publ 215, IAHS Press, Wallingford pp 329–338
Walling DE, Owens PN, Leeks GJL (1999) Fingerprinting suspended sediment sources in the catchment of the River Ouse, Yorkshire, UK. Hydrol Process 13:955–975
Wilkinson SN, Hancock GJ, Bartley R, Hawdon AA, Keen RJ (2012) Using sediment tracing to assess processes and spatial patterns of erosion in grazed rangelands, Burdekin River basin, Australia. Agr Ecosyst Environ 180:90–102
Yaalon DH (1997) Soils in the Mediterranean region: what makes them different? Catena 28:157–169

Yassoglou NJ, Kosmas C (1997) Desertification in the Mediterranean Europe. A case in Greece. In: RALA (ed) Rangeland desertification report, Report N°200. RALA Ed, Reykjavik, Iceland, pp 27–33

Yeager KM, Santschi PH, Phillips JD, Herbert BE (2005) Suspended sediment sources and tributary effects in the lower reaches of a coastal plain stream as indicated by radionuclides, Loco Bayou, Texas. Environ Geol 47:382–395

Yu L, Oldfield F (1989) A multivariate mixing model for identifying sediment source from magnetic measurements. Quat Res 32:168–181

Chapter 6
Wrapping Up Findings in a Sediment Transport Model for Wadi Al-Arab

6.1 Introduction

To encounter water scarcity, Jordan contemplated and partly implemented different water management strategies to harvest rainwater and runoff, and artificially recharge ground water aquifers. The methods include rain water harvesting in cisterns (Abdulla 2006), construction of water reservoirs (Obeidat et al. 2011), and other methods of Managed Aquifer Recharge (MAR) as spreading methods (infiltration basins, irrigation, etc., MWI 2001) or in channel modifications as check dams or leaky dams with recharge releases (Wolf et al. 2007). The good effectiveness of check dams has been documented by several authors for the Arab world and east Asia and a compilation of findings with up to 40 % recharge increase is listed in Renganayaki et al. (2013). Other advantages include the reduction of evaporation loss due to subsurface water storage, the enhancements of ground water quality when fed by rainwater and runoff, and impeding erosion through the reduction of runoff velocities. However, in areas like Jordan where rivers transport high sediment yield especially of fine grain sizes <63 μm, dams enhance sedimentation and siltation, which decreases infiltration with time and may even contaminate a water body with sediment bound pollutants (Ghrefat and Yusuf 2006; El-Radaideh 2010; Al-Ansari and Shatnawi 2011; Al-Ansari and Knutson 2012a; Renganayaki et al. 2013; Kraushaar et al. 2014). Thus, the planning of in channel modifications to enhance infiltration and groundwater recharge should carefully evaluate suitable areas following the following criteria:

1. Low sediment yield in the river to guarantee a long life span of the constructions and a high efficiency of the investments.
2. Low sediment connectivity of the catchment's hillslopes to the river. This will assure the sediment yield robustness to land use changes (agricultural origin, road construction etc.) on the hillslopes.
3. Preferentially located in headwaters as discharge volume and velocities of flash floods have stronger impacts on the endurance of constructions downstream.

S. Kraushaar, *Soil Erosion and Sediment Flux in Northern Jordan*,
Springer Theses, DOI 10.1007/978-3-319-31888-2_6

4. Preferentially not located downstream of a Waste Water treatment plant as effluents may not suffice quality standards throughout the year and pollute the aquifer.
5. Not located in headwaters of springs, so that the infiltrated water will be stored in the aquifer and not yielded directly.

In Wadi Al-Arab MAR structures are contemplated and thus the spatial assessment of erosion hot spots and sediment yield of interest. This chapter will use the findings from previous ones (Chaps. 3, 4 and 5) and merge the results and knowledge in the SedNet model. The model allows combining measured landscape parameters and physical reasoning of erosion and deposition factors, to recursively test the synthesis of soil erosion and sediment transport in the catchment (Wilkinson et al. 2004). This will help to evaluate the gained knowledge and understand the linkages between upstream sources and downstream impacts on a catchment scale.

The SedNet model is a sediment budgeting model that couples the conceptual erosion model RUSLE (Renard et al. 1997) with a physically-based transport model (Wilkinson et al. 2004). It can predict sediment loads from different sources such as hillslopes, gullies and riverbank as well as their deposition on the floodplains, as bedload or in reservoirs (de Rose et al. 2005). The model was developed by the Australian National Water Resource Audit (Prosser et al. 2001) to suit Mediterranean to semi-arid regions in Australia and applied to macro scale catchments of >1000 km^2. Nevertheless, in the past it was successfully implemented on a scale of 100–1000 km^2, similar to the scale of the catchment under focus in this work (Rustomji et al. 2008). SedNet is generally used to identify high erosive areas, evaluate sediment yield, and spatial patterns of sediment sources and deposition in a catchment that needs prioritized rehabilitation efforts. The model allows the generation of future scenarios to support management decisions (McKergow et al. 2005). The RUSLE model is based on a simple factor multiplication, unable to take feedback calculations into account. However, for the Wadi Al-Arab catchment in northern Jordan this method is suitable for several reasons:

- Physically based erosion models are mostly developed for small scale, humid regions (<10 km^2) and demand numerous parameters. Many of them are not even measurable and have to be calibrated by observed data, adding to the uncertainty of the model (Merrit 2003). Additionally mathematical assumptions that are implied to physical based models developed for humid regions, might not apply to Mediterranean or semi-arid areas, which make the transference problematic.
- Countries like Jordan lack on comprehensive and high resolution maps, such as parameterized soil, geology or even topographic maps. Often weather stations are scarce and continuous measurements do not exist.
- The RUSLE has the biggest empirical background data set with more than 10,000 plot years of runoff and soil loss data (Wischmeier 1978). It is suitable for catchments with limited data sets and where average soil loss rates per year are sufficient for the research question (Panagos et al. 2012, 2014). Thus, it does not imitate a precision that is not integer concerning the data scarcity. Moreover,

a more complex physically based model does not necessarily perform better due to the numerous input parameters and their uncertainties (Jetten et al. 2003). Nevertheless, it still produces valuable information to allocate erosion hot spots and plan mitigation strategies (Meusburger et al. 2010).

Hence, the use of the conceptual model RUSLE for the estimation of soil erosion on the slopes in combination with a physical based transport model is viewed as the ideal compromise for the analysis of soil erosion and sediment transport in Wadi Al-Arab.

This chapter describes the parameterization of relevant raster maps for the setup of the SedNet model in a data scarce catchment in Wadi Al-Arab/northern Jordan. Field measurements, such as the average yearly sedimentation rate in the reservoir, the mean yearly erosion values of the three relevant land use units (olive orchards, agricultural fields and vegetated slopes), and ^{137}Cs analysis of terrace profiles are used for calibration and validation of the model. Further developments of the model calibration include:

- The inclusion of aspect dependent differences in soil and vegetation cover in the C factor,
- The modeling of the hillslope deliver ratio (HSDR) with satellite data and slope information.

Results allow a careful assessment of suitable areas for the construction of check or leaky dams in the Wadi bed.

6.2 Model Selection and Description—SedNet

A detailed description of SedNet is given in Prosser et al. (2001), de Rose et al. (2005), Wilkinson et al. (2004, 2006, 2009), and others. This chapter will just give a brief summary of the relevant aspects of the model.

SedNet builds up on a river network derived from the hydrological analysis of a given DEM. The reaches between the river junctions each define a sub-catchment. In case of Wadi Al-Arab 17 sub-catchments with an area of minimum 1–33 km^2 each were generated.

From the eroded hillslope only a defined part of the sediment amounts calculated with the conceptual erosion model RUSLE (Renard et al. 1997) reaches the riverbed and is included into the suspended sediment transport. This connectivity is expressed via the hillslope delivery ratio. Riverbank erosion additionally occurs according to discharge and riverbed characteristics, such as slope, riparian vegetation and floodplain width. Highest bank erosion is expected in links with wide alluvial deposits, steep slopes and no riparian vegetation. Gullies contribute to the sediment load according to their extent in the catchment (Wilkinson et al. 2008; Fig. 6.1).

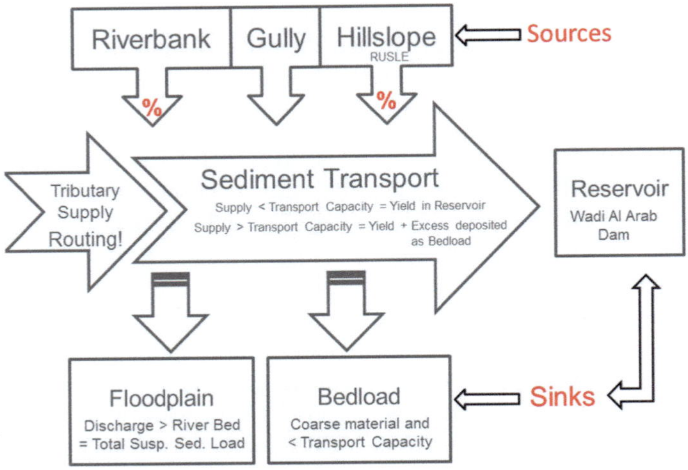

Fig. 6.1 Schematic diagram of the SedNet model

Concerning the sediment transport in the riverbed the SedNet model differentiates between fine (<63 μm) in suspension and coarse sediment fractions (≥63 μm). Suspended sediments are assumed to be transported through the river network by the SedNet model. However, if an overbank flow event is computed, fine grained material is deposited in the floodplains, depending on sediment settling rate and residence time of floodwaters. The rest of the sediments is routed through the system until the reservoir. Deposition in the reservoir follows the empirical relation by Brune (1953).

For coarse sediments the sediment transport capacity (derived from the geometry of the riverbed, discharge quantity and velocity, roughness) is the critical threshold for further mobility. In the case the sediment load exceeds the transport capacity; coarse sediments are deposited as bedload. In case the transport capacity exceeds the sediment supply, the coarse sediments are routed through the network until the reservoir. Usually rivers in Mediterranean regions have a high transport capacity and the sediment yield is often only limited by sediment delivery from the slopes (Vanmaercke et al. 2011). Thus, sediment deposits in the riverbed are often negligible.

6.3 Methods

6.3.1 Field and Laboratory Work for Model Setup

Data necessary for the model parameterization were gathered during several field campaigns between 2010 and 2013.

Special focus was set on the generation of the gully density grid, the development of a parameterized genetic soil map, the hillslope delivery ratio and climate grids as those values have been reported to be sensitive when considering total sediment export from the model (Fentie et al. 2005). An overview of the input parameters of the SedNet model and the data procession is given in Fig. 6.2.

	Model Parameters	Data Procession
Basic Data	DEM (29x29)	Aster GDEM V2 (product of METI and NASA).
	Lake reservoir	Echo sounding and GIS analysis of sedimentation in the Wadi Al-Arab reservoir (October 2012).
	Gully density grid	Gullies in 10% of the catchment's area (randomly chosen) mapped and measured, interpolated according to bedrock characteristics (November 2012).
	Floodplain grid Riperian Veg. grid	Mapped throughout the Wadi (June 2013).
Hillslope erosion grids	C-factor: Cover	Calculated from empirical relationships (Agricultural Handbook 537); aspect dependencies included.
	R-factor: Rainfall erositivity	Interpolated from published values of 6 climate stations (Eltaif et al., 2010).
	K-factor: Soil erodibility	Genetic soil map derived from 28 recorded soil profiles following the Scorpan approach (Jenny, 1994).
	S-factor: Slope steepness L-factor: Slope length	C++ analysis following van Remortel et al. (2004).
	Hillslope Delivery Ratio	Computed using slope and vegetation cover (NDVI) (Rustomji, 2010).
	Landuse map	Based on two Landsat ETM+ GLS 2010 scenes and a hierarchical classification analysis (Mallast 2011, modified Kraushaar, 2012).
Weather Climate and Hydrology Data	Mean air temperature grid	Daily records from 6 climate stations closest to the catchment (MWI, 2012); 1968-2007 averaged, ordinary kriging.
	Mean annual rainfall grid	Daily records from 13 rain stations in, or 2.5 km around the catchment (MWI, 2012); 1938-2007 averaged, ordinary kriging.
	PET-Rainfall Ratio grid	PET calculation: Daily records from 5 climate stations closest to the catchment (MWI, 2012); 1968-2007 averaged, calculated following Penman Monteith, ordinary kriging.
	Daily Flow data (tabular)	Daily inflow data from the Jordan Dam Authority (1987-2012).
Nutrient Datasets and point sources	Soil clay percent grid	Genetic soil map derived from 28 recorded soil profiles following the Scorpan approach (Jenny, 1994).
	Nitrogen and Phosphorus grid	Genetic soil map derived from 28 recorded soil profiles following the Scorpan approach (Jenny, 1994).
	Dissolved nutrient point sources (tabular)	3 discharge measurements: Dissolved inorganic and organic nitrogen and phosphorus.
	Dissolved runoff nutrient concentration (tabular)	Runoff data from 4 event measurements: Dissolved inorganic and organic nitrogen and phosphorus.

Fig. 6.2 Model parameters for SedNet

6.3.2 Gully Density Map

For the gully density mapping 20 points with 1 km buffer (=3.14 km^2) were randomly spread over the catchment (Fig. 6.3) covering areas of different bedrock with more or less marly substrate. Eight of these defined areas (=25.12 km^2 = 10 % of the catchment) were investigated for noticeable structures that resembled linear erosion features using aerial pictures. A ground truthing campaign was implemented in five areas (=15.7 km^2 = 6 % of the catchment) and all erosion features

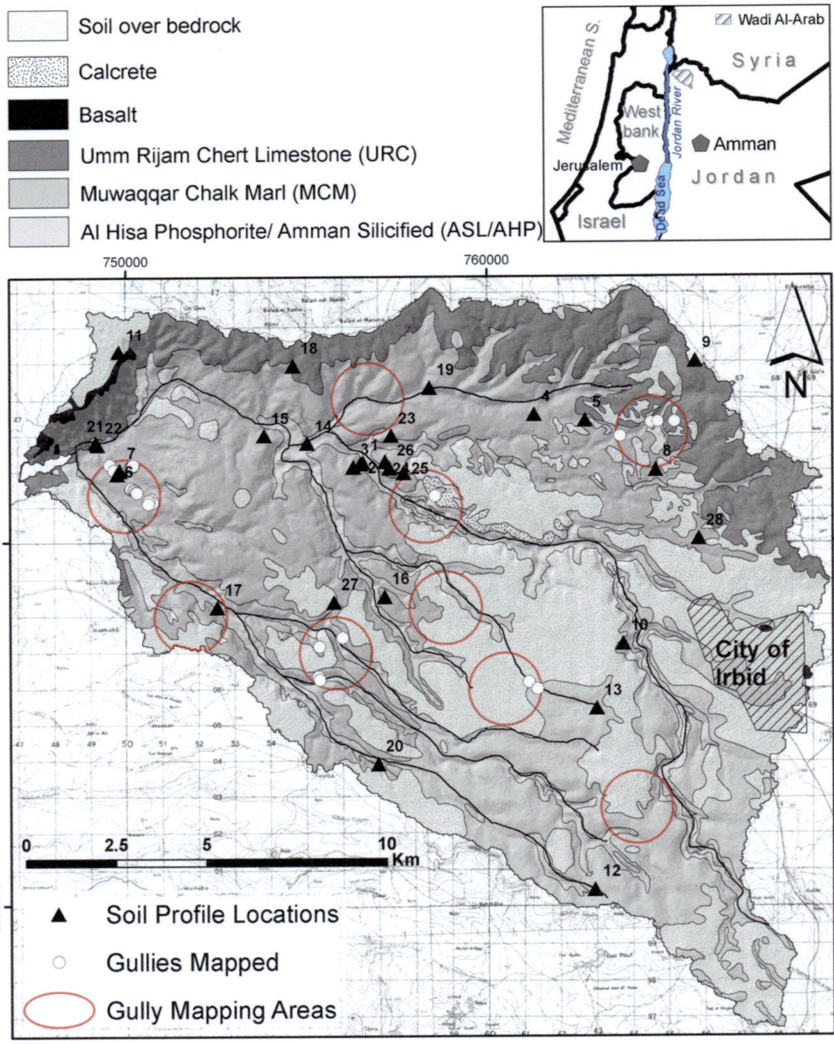

Fig. 6.3 Major geological units, soil profile locations and mapped gullies in Wadi Al Arab (Projection UTM, WGS84, zone 36 N)

mapped and measured. Results showed that the aerial photo interpretation is useful to spot gullies and define their length, however a good transect estimation of the gully course was not possible using aerial photos alone. Summarized gully length divided by the area was recorded for marly geology (MCM and URC: 3.94 km = 0.49 km km^{-2}; Fig. 6.3) and hard limestone (ASL/AHP: 0.78 km = 0.05 km km^{-2}; Fig. 6.3).

6.3.3 *Pedogenetic Soil Map*

Several input parameters of the SedNet model demanded spatial information on grain size fractions, bulk density, Kf-values, total organic carbon, nitrogen, and phosphorus of the upper soil layers. To meet these demands 28 profiles were recorded following the catena principle (Milne 1936) in each of the major geological units ASL/AHP, MCM, URC, and basalt. In respect to the exposition differences, an aspect differentiation was included and profiles along north and south facing slopes opened. Up to 3 profiles were dug on each of the different morphological positions including upslope, elongated slope, saddle position, foot slope with and without colluvium, and alluvial positions on different exposed slopes.

Using the German genetic classification approach (Ad-hoc-Arbeitsgruppe (AG) Boden 2005) horizons were identified and mixed soil samples collected for further analysis of the above mentioned parameters. Each profile represents a certain morphological position on the specific geological unit, with a certain aspect, and under different land use (=landscape unit). Infiltration test using a double ring infiltrometer were performed. These units were then spatially constrained in the DEM with Arc Map 10.1 following a decision tree that relies on flow accumulation, profile curvature, slope, aspect, and land use rasters (Fig. 6.4a, b).

6.3.4 *The Hill Slope Delivery Ratio*

The hill slope delivery ratio (HSDR) was modelled assuming that slope and vegetation cover are the driving factors of sediment delivery on the slope following Rustomji et al. (2010).

$$\mathrm{HSDR}_{\mathrm{unscaled}} = \left(1 - \frac{\mathrm{Slope}_{\mathrm{max}} - \mathrm{Slope}_{\mathrm{i}}}{\mathrm{Slope}_{\mathrm{max}}}\right) * \mathrm{w}_1 + (1 - \mathrm{Cover}_{\mathrm{i}}) * \mathrm{w}_2 \qquad (6.1)$$

where $\mathrm{HSDR}_{\mathrm{unscaled}}$ is the hill slope delivery ratio unscaled for each sub-catchment; $\mathrm{Slope}_{\mathrm{max}}$ = maximum slope for the each sub-catchment (%); $\mathrm{Slope}_{\mathrm{i}}$ = the slope of $\mathrm{pixel}_{\mathrm{i}}$; $\mathrm{Cover}_{\mathrm{i}}$ = reclassified NDVI values for $\mathrm{pixel}_{\mathrm{i}}$; and w_1 and w_2 as relative weights for each variable ($\sum w = 1$).

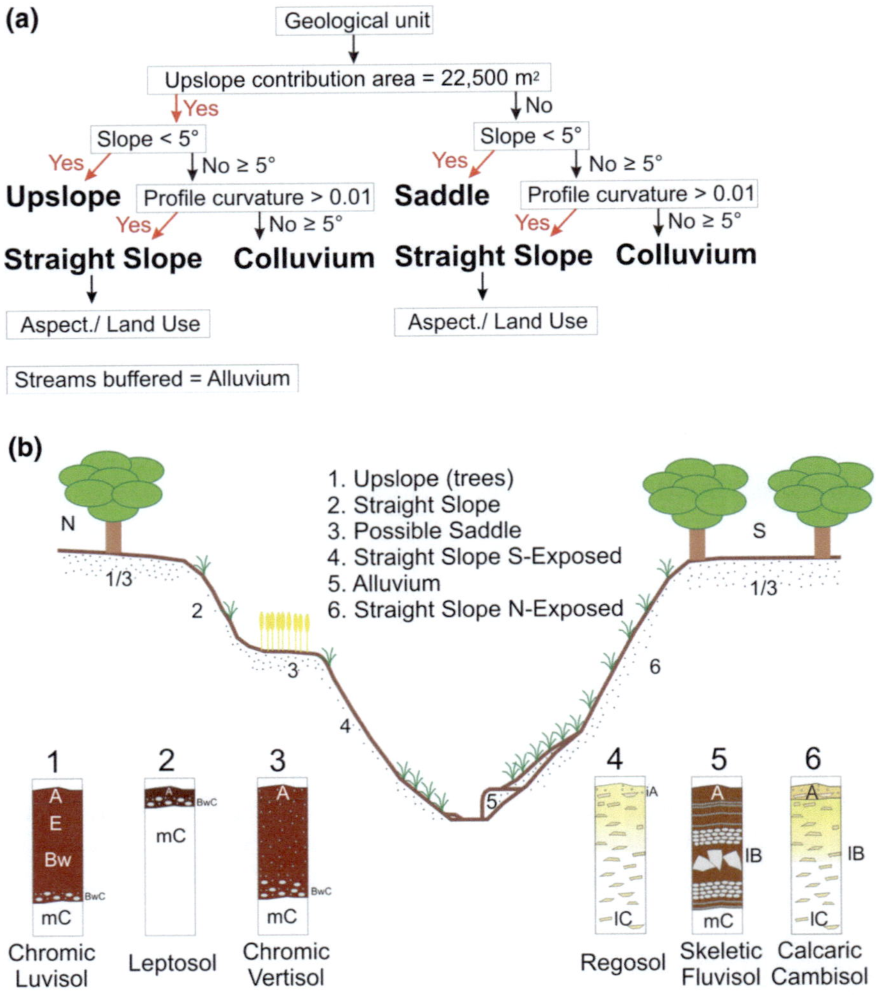

Fig. 6.4 Compilation of a regionalized soil map following the Scorpan approach (Jenny 1994). **a** Identification of landscape units. **b** Standardized Catena for the MCM unit

The NDVI is derived from spectral reflectance measurements acquired in the visible (red) and near-infrared regions of the two Landsat ETM + GLS 2010 scenes. Following Rustomji et al. (2010) the relative weights were assigned as as equal ($w_1 = w_2 = 0.5$).

The assumption is then made that the sub-catchment wide $HSDR_{unscaled}$ mean ($\overline{HSDR_{unscaledSC}}$) should equal an iterative modelled HSDR ($\overline{HSDR_{mod}}$). In this case a $\overline{HSDR_{mod}} = 0.135$ gave the best fit of the model to the yearly expected sediment yield. For each pixel the HSDR is then scaled ($HSDR_{scaled}$) following Eq. (6.2).

$$HSDR_{scaled} = HSDR_{unscaled} * \frac{\overline{HSDR_{mod}}}{\overline{HSDR_{unscaledSC}}} \tag{6.2}$$

6.3.5 *Field Data as Quality Threshold for Model Performance*

The model is calibrated on the total yearly sediment load that reaches the dam (Chap. 4) and validated with the results of the erosion measurements from (Sects. 3.1, 3.2 and 3.3).

6.4 Calibration, Validation and Sensitivity Analysis of the SedNet Model with Field Results

6.4.1 *Calibration of the SedNet Model*

The SedNet model is stepwise parameterized and calibrated against the total sediment load measurements from the Wadi Al-Arab reservoir (=100,125 t year^{-1} ± 6 %).

Figure 6.5 shows enhancements in model results:

1. Default set up of the model with the parameterized raster information.
2. Added regional information on the gully transects, riverbed morphology, and sediment characteristics (bulk density, grain size fractions).
3. Regionalized runoff coefficient.
4. Included aspect differentiation in the cover factor.
5. The modelled hillslope delivery ratio (Sect. 6.4).

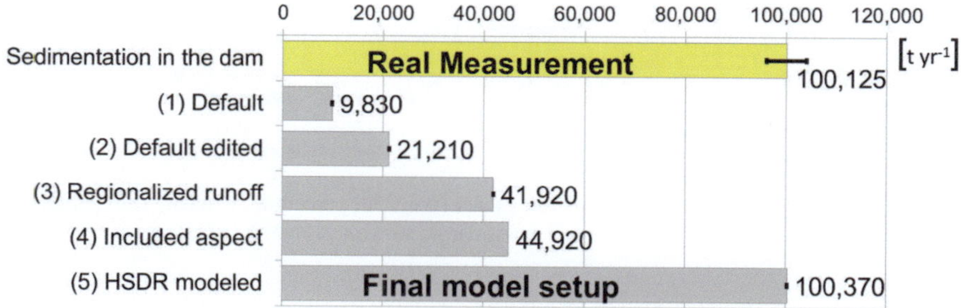

Fig. 6.5 Stepwise model calibration with yearly reservoir sedimentation rate

The final model setup is done according to (5) in Fig. 6.5 with a deviance to the measured sediment load of +0.24 %.

6.4.2 *Validation of the SedNet Model with Erosion Measurements*

The RUSLE results of the SedNet model display the possible translocation of material on the slope per pixel and thus are compared to the measured values (Fig. 6.6). Results of the erosion measurements identified olive orchards on steep slopes (>10 %) having the highest erosion potential in the region with 95 ± 8 t ha^{-1} $year^{-1}$ followed by natural vegetated slopes with 40.2 ± 2.6 t ha^{-1} $year^{-1}$ of dislocated material and fields with 1.22 ± 0.06 t ha^{-1} $year^{-1}$ sediment yield. The comparison shows a good average fit between agricultural fields and slopes modelled. The measurements under the olive orchards indicate much higher average erosion rates (99 t ha^{-1} $year^{-1}$) than the modelled ones (46 t ha^{-1} $year^{-1}$). Due to the fact that the olive mound measurement method was only possible in orchards on steep slopes with strong erosion signs, as described in Sect. 3.2, the method can only display these extreme values. Thus, the highest measured erosion values are in agreement with the highest modelled outliers and prove the model to give reasonable average erosion values.

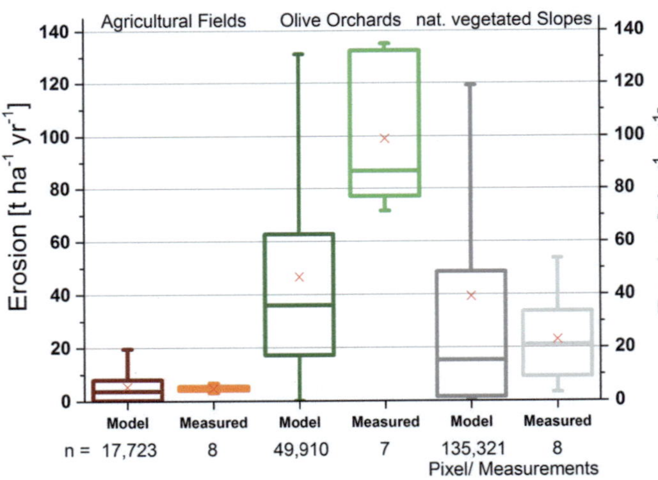

Fig. 6.6 Measured versus modelled erosion values (*box* 25th percentile, median, 75th percentile; whiskers: outliers—75th Percentile + (1.5 Interquartile range); *X* = arithmetic mean)

6.4.3 Sensitivity Analysis of the SedNet Model

Following Newham et al. (2003) and Fentie et al. (2005) the SedNet model was evaluated using a simple and local one-at-a-time sensitivity analysis. 17 scenario parameters and 10 hydrologic parameters were alternated by ±10 and ±5 % from their baseline values and a dimensionless sensitivity index (S) calculated with the varied results of the total sediment deposition in the reservoir (Eq. 6.3).

$$S = \frac{\partial X}{\partial Y} \tag{6.3}$$

where ∂X is the relative change of model results and ∂Y the relative change in parameter from the baseline (Fentie et al. 2005).

The scenario parameter with the highest sensitivity index (close to 0.83) is the hillsope delivery ratio. The gully average cross section area, the gully erosion reduction factor, years of elevated erosion, sediment bulk density, the floodplain settling velocity, the proportion of suspended material (<63 μm) and bed material (>63 μm <2 mm) showed a *S*–value below 0.2 and thus indicate that the model is not highly sensitive to them. All of these values were derived from field observation as described previously.

The hydrological parameters over all showed a sensitivity index with maximum values of 0.6 (=reoccurrence interval), followed by the model calculated parameters bankfull discharge and median overbank flow (S ~ 0.2).

In regard to the general higher sensitivity index of the hydrological parameters a comparison of the discharge and runoff curve numbers between model and measurements was implemented (Fig. 6.7). The estimated runoff curve number by the model (0.0105) agrees to the ones independently generated (0.016 ± 0.082) by

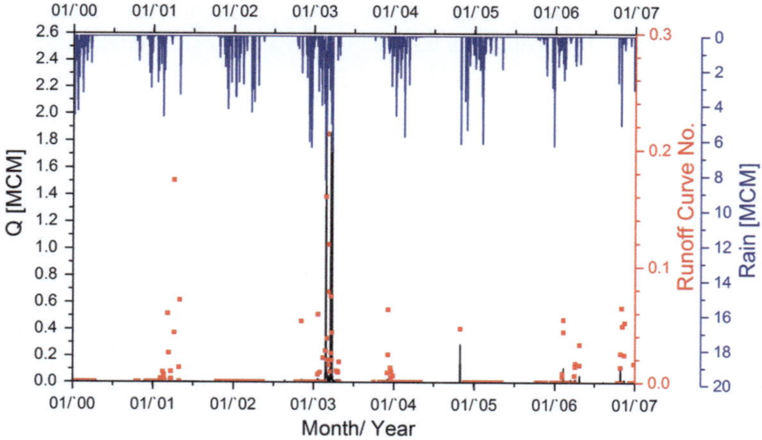

Fig. 6.7 Modelled precipitation for the Wadi Al-Arab plotted against discharge (Q) and the runoff curve number for the rainy season 01/2000 until 02/2007

averaged precipitation values calculated via JAMS (Jena Adaptable Modelling System; Kralisch and Krause 2006), divided by the measured discharge (JDA data 2011). The model calculates a regionalized runoff following Wilkinson et al. (2006). In case of Wadi Al-Arab the regionalization relies on the relationship of the measurements from two gauging stations plotted against their aridity index (=*ai*; Eq. 6.4) as described by Wilkinson et al. (2006).

$$ai = \frac{PET}{P} \tag{6.4}$$

where *PET* is the potential evapotranspiration and *P* is the precipitation at the location of the gauging stations.

These two gauging stations, however, lie very close to each other at the entrance of the reservoir in the west of the catchment where the *ai*(=3.6) is the highest. This relationship cannot give a good fit for the eastern part (*ai* = 2.7). Hence, the modeled average flow is estimated by the model to result in 708 × 10^3 m^3 per year which deviates by 14.21 % from the measured average discharge values from the hydrological year 1987–2012 of 620 × 10^3 m^3. The year 2003 was an exceptional rainy year with over 10 × 10^6 m^3 and was not included into the average discharge calculations.

6.5 Results and Discussion

6.5.1 Sediment Budget Results

In the sediment budget of the Wadi Al-Arab catchment (Table 6.1) the gross input of sediments to the system derives from hillslope erosion with 108 × 10^3 t $year^{-1}$ followed by Gully sediments with 24 × 10^3 t $year^{-1}$. Bank erosion plays only a minor part in the budget with less than 1 × 10^3 t $year^{-1}$ of predicted input load.

Table 6.1 Sediment budget model results for Wadi Al-Arab

Suspended sediments	Input (10^3 t $year^{-1}$)	Output (10^3 t $year^{-1}$)	
Hillslope	108	Reservoir deposition	100.36
Gully	21.96	Floodplain deposition	23.17
Bank	0.14	Export from network	6.57
Bedload sediment	Input (10^3 t $year^{-1}$)	Output (10^3 t $year^{-1}$)	
Gully	2.02	Reservoir deposition	0.01
Bank	0.01	Channel deposition	1.99
		Export from network	0.03
Total	132.13		132.13

Fig. 6.8 Wadi bed: **a** Confined river bed in headwaters, **b** high terraces 5–8 km in front of the dam, widening of the river bed, **c** floodplain, shortly before the inlet to the Wadi Al-Arab reservoir

Around 80 % of the sediment load that is supplied to the river is routed through the system and sedimented in the reservoir. Only 18.4 % feeds the floodplains and 1.6 % remains as bed load in the Wadi, predominantly sedimented at the entrance of the reservoir. The predicted floodplain deposition rate ranges between 0 and 20 mm with an average of 10 mm.

The high amount of sediments that reach the dam and the low bed load modelled correlates with the field observations of a small frequency meandering and gravely riverbed confined by steep slopes. Only small areas of sediment deposition in shallow convex banks exist in the major part of the Wadi. 5–8 km in front of the reservoir, the riverbed widens from measured 2 to 8–14 m and 6 m high paleo terraces and foot slope positions become visible alternating at the sides of the Wadi (Sect. 2.3; Fig. 6.8). Analysis of 38 reservoir and two terrace samples demonstrated that all of the reservoir samples exhibit a ^{137}Cs content similar to the surface samples from the catchment and hence, had surface contact between 1954 and 2012. Contrastingly the terrace sediments were ^{137}Cs free. The missing ^{137}Cs and the enhanced reddish color (Munsell value of 7.5 YR 4/4) on top of the terrace profiles, indicate that they have been geomorphological stable at least for the past 60 year. Thus, they do not serve as potential temporary storage for recent fluvial transported sediments, nor as sediment source for the dam sediments (Chap. 4).

These geochemical results as well as the observations in the Wadi bed point to a high sediment connectivity between the different compartments slope, Wadi bed and reservoir following Fryers (2013). A good water mediated transfer is therefore given which corresponds to the model results as listed in Table 6.1.

6.5.2 *Patterns of Erosion in the Wadi Al-Arab Catchment*

The sediment sources hillslopes, gullies, and river banks show different erosion patterns throughout the catchment (Fig. 6.9). In general the raster map derived from the RUSLE results identifies areas of high hillslope erosion probability per pixel. These areas concentrate on south facing slopes in the north, northwest and sub-catchment 15 and correspond with regions of higher relief energy in the north, northwest and south of the catchment. The coherency of the eroded flanks with the patterns of high erosion probability in the map is visible when compared to the satellite image (Fig. 6.9a, b). Cerdá (1998) and Bochet and García-Fayos (2004) report for Spain that higher erosion values are found on south facing slopes as they have less protective vegetation cover and hence, less organic content in the top soil layer (first 2 cm), which favors higher runoff values and less soil structure stability. Cerdà (1998) proved for bare soils on south facing slopes higher bulk density values, less soil depth and crusting as responsible characteristics for enhanced runoff and erosion processes.

Figure 6.9c, d show hillslope, gully and bank erosion values in t km^{-2} $year^{-1}$ to define areas of accelerated erosion compared to the catchment size, and hence, support the allocation of prioritized areas for mitigation strategies. The high hillslope erosion rates are mainly existent in the north, north- west and south of the catchment (sub-catchments 2, 7, 8, 13, 18, 16; Fig. 6.9c) and assist the results of the erosion probability map.

Gully erosion concentrates on the sub-catchments in the north and north-west (sub-catchments 1, 2, 3, 7, 8, 9, 13, 18) and becomes more moderate to the south and east (Fig. 6.9d). This relates dominantly to the underlying unconsolidated marly geology (MCM; Fig. 6.3) paired with the higher relief energy in the north and north west. Gullying in marly landscapes of the Mediterranean is reported in various studies (Kosmas et al. 1997; Casalí et al. 1999; Cerdà 2002; Faust and Schmidt 2009; García-Ruiz 2010; 2013). Reasons are physical properties as texture and impermeable layers, geochemical properties as the high sodium saturation that leads to the dispersion of soil aggregates, and the preferential agricultural use of these areas due to the high crop yield (Faust and Schmidt 2009). Gullies have the disadvantage of direct soil loss and serve additionally as effective links routing sediments from the surrounding immediately towards the Wadi bed (Faust and Schmidt 2009) and thus, guarantee a good sediment connectivity of hillslopes and river bed as described above.

6.5.3 Next Page

Bank erosion occurs intensively west of the dam (sub-catchment 1), due to the steep slopes. Furthermore, it exists stronger in the main river bed stretches before the dam, resulting from the higher transport capacity downstream (sub-catchments 2, 13; Fig. 6.9d).

Total erosion rates of all possible sources are the highest in the north and northwest of the catchment. The general erosion pattern follows the underlying geology and delineates the marly MCM from the silified ASL/AHP (Fig. 6.9e).

The total sediment yield describes the supply minus the depositions and relates to the connectivity of the hillslopes to the river network. Sub-catchments 9, 14 and 17 indicate high delivery, although total erosion rates are moderate. This results from the previous discussed high gully erosion values that route the sediments directly to the river bed and the V-shaped valleys in the Upper part of the Wadi,

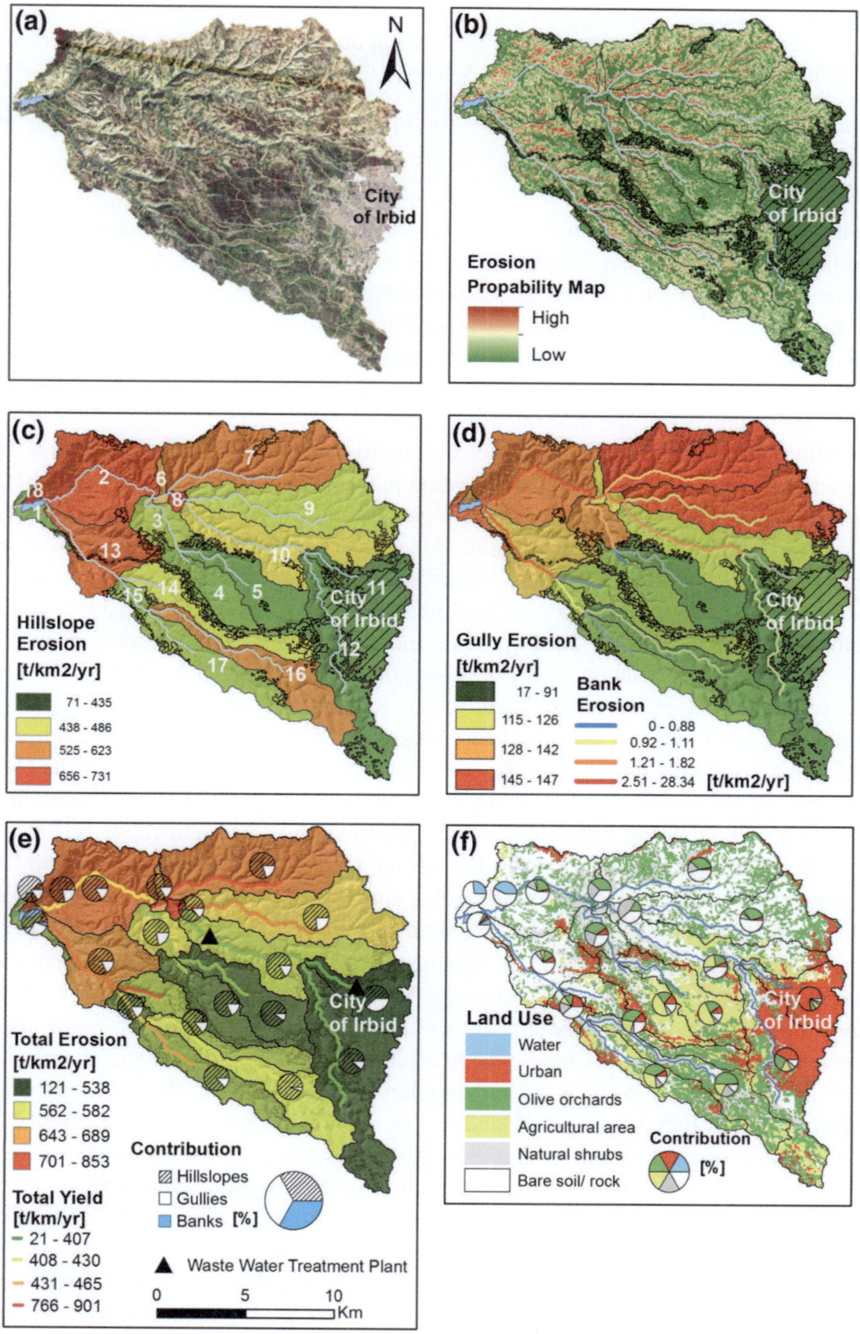
(a)
N
City of Irbid
(b)
City of Irbid
Erosion Propability Map
High
Low
(c)
Hillslope Erosion [t/km2/yr]
71 - 435
438 - 486
525 - 623
656 - 731
City of Irbid
(d)
Gully Erosion [t/km2/yr]
17 - 91
115 - 126
128 - 142
145 - 147
Bank Erosion
0 - 0.88
0.92 - 1.11
1.21 - 1.82
2.51 - 28.34 [t/km2/yr]
City of Irbid
(e)
Total Erosion [t/km2/yr]
121 - 538
562 - 582
643 - 689
701 - 853
Total Yield [t/km/yr]
21 - 407
408 - 430
431 - 465
766 - 901
Contribution
Hillslopes
Gullies
Banks [%]
Waste Water Treatment Plant
0 5 10 Km
City of Irbid
(f)
Land Use
Water
Urban
Olive orchards
Agricultural area
Natural shrubs
Bare soil/ rock
Contribution [%]
City of Irbid

◀ **Fig. 6.9** Erosion and sediment yield predictions in the Wadi Al-Arab catchment, numbers in Figure **c** are valid for the figures **c–f**. **a** Aerial picture of the catchment (Esri base map), **b** erosion probability raster map of the RUSLE results, **c** hillslope erosion (t km^{-2} $year^{-1}$) per sub-catchment, **d** gully and bank erosion (t km^{-2} $year^{-1}$), **e** total erosion (t km^{-2} $year^{-1}$), total yield (t km^{-2} $year^{-1}$) and the contribution (%) of each source, **f** land use map with land units percentage in diagrams

both being responsible for a high connectivity between the compartments. In the northwest, sub-catchment 2 confirms with high erosion rates per square kilometer but less connectivity the field observations of high paleo-terraces that flank the sides of the Wadi acting as intermediate storage for sediments from the slope, not from the river. In these areas the system is less sensible to changes on the slope, although high erosion rates exist, whereas the water mediated transfer of sediments from the east and southwest of the catchment is higher.

Although sediment yield results correspond to qualitative observation it has to be stated that due to the lack of gauging station in the catchment the model shows high deviations (14.21 %) in the average flow generation. This results in lower transport capacities and higher rates of possible deposition or lower rates of bank erosion, respectively. Previously discussed results have to be reviewed in respect to that.

The relative contribution of each source (hillslope, gullies, river banks) to the sediment yield is shown in Fig. 6.9e, indicating that in the majority of the sub-catchments the relation of hillslope to gully sediments is 3:1, as mentioned in the sediment budget. A general trend exists to less gully sediments coming from south-east and therefore, the silicified background with less relief energy with one exception—sub-catchment 11 in the east is covered by the urban sprawl of the City of Irbid and shows 1/3 of the total amount contributed by gully erosion. The sealing of soil prohibits strong hillslope erosion. Additionally, roofs and roads accelerate runoff concentration, which cause increasing rill and gully erosion as reported before by Nyssen et al. (2002), Ehiorobo and Audu (2012), and Valentin et al. (2005). Bank material only contributes significantly to the yield in sub-catchment 1.

Suspended Sediments from the hillslopes are the major source of eroded and transported sediments and derive mostly from olive orchards and naturally vegetated but grazed slopes. The higher the amount of urban areas or fields in a sub-catchment the less hillslope erosion it shows. The erosive land use classes are stronger represented in areas of higher relief energy to the north and north west which makes it difficult to say if land use or declination are primarily responsible for the high erosion values (Fig. 6.9f). When modeling a scenario with adapted olive farming in the same mountainous relief, including older trees with a higher canopy cover (40 % instead of 20 %) and reduced tillage leaving 60 % (instead of 5 %) mulch cover throughout the rainy season the hillslope erosion rate would decrease by 48 % to 55 × 10^3 t $year^{-1}$. The overall annual sediment load in the dam would be reduced by 54 % to 60 × 10^3 t $year^{-1}$ and the floodplain deposition rate would decline by 43 % resulting in 13 × 10^3 t $year^{-1}$. The SedNet model reacts very sensitive to this management change in the conventionally tilled olive orchards, which cover currently 27 % of the catchment.

6.5.4 *Suitable Locations for in Channel Modifications for Artificial Aquifer Recharge*

Intended check dams installations in the Wadi Al-Arab river bed should concentrate on head waters in sub-catchments with low total erosion rates and low sediment connectivity. The location downstream of Waste Water Treatment Plants (2, 6, 8, 10, 11) should be possibly excluded. The enhanced infiltration of effluents, which have been observed being discharged numerous times in the field, might be of harm to the ground water (Fig. 6.9e). Springs often exist as contact springs in the north west of the Wadi between the B3 and B4 units (Siebert et al. 2014). Most of them discharge around 50 m above the river bed and therefore have no impact on the selection of suitable areas in the riverbed. Hence, preferential sub-catchments for check dams are 4, 5, 12, as they unite all the criterias best. Sub-catchments that show high erosion values coupled with high connectivity and therefore, high sediment yield (7, 8, 13), should be avoided especially since the high sediment load will reach the watercourse in short time. Additionally, sub-catchments even with moderate erosion values and a high connectivity/sediment yield (14, 15, 17) are advised against as the system is highly sensitive to changes on the slope. Thus, land use and/or management alterations, as well as road construction could trigger higher erosion values. In any case these technical considerations should go hand in hand with an integrated management plan for the increasing grazing activities on the slopes as well as the conventionally tillage under growing olive orchards as these are the most prominent sediment sources in the system which need to be optimized when aiming at a long lasting sustainable ground water recharge constructions.

6.6 Conclusions

The Sediment River Network model (SedNet) was parameterized for the Wadi Al-Arab catchment in northwestern Jordan. Therefore, detailed soil and gully mapping was implemented. This appears to be especially important as a simple sensitivity analysis identified sediment bulk density, certain grain size fractions as well as gully morphology as sensible scenario parameters. The hillslope sediment delivery ratio was modelled following Rustomji et al. (2010) with reclassified NDVI values and declination. The model was calibrated with the average yearly sediment deposition rate of the Wadi Al-Arab reservoir and validated using several erosion measurements from the major land use units olive orchards, natural vegetated slopes and agricultural areas. Results show that the model is capable to predict erosion on the slope well, whereas the transport of sediments underlies a high uncertainty due limited gauging stations in the catchment. The reason for the mean differences of the olive orchard erosion values lie in the nature of the mound measurement method, used for the calculation of medium-term average erosion rates. The method is only suitable for steep slopes >5°–6° (Kraushaar et al. 2014) and therefore correspond correctly only to the highest modeled erosion rates. Final erosion results show that especially the north, northwest and west of the catchment suffers from high erosion values, mainly from hillslopes. Highest erosion values correspond to areas of high relief energy, south exposed, under cultivation of olives or used for grazing, and with the marly unit MCM as geological background.

Connectivity is especially high in the north and west and thus, the total sediment yield particularly sensitive to changes of land use in the sub-catchment. Areas suitable best for check dams in the Wadi bed are head waters in the east and middle of the catchment.

References

Abdulla FA, Al-Shareef AW (2006) Assessment of rainwater roof harvesting systems for household water supply in Jordan. In: Hlavinek P Kukharchyk T Marsalek J Mahrikova I (eds) (2006) Integrated urban water resources management. Proceedings of the NATO advanced study institute on integrated urban water resources management, held in Senec, Slovak Republic, 19–23 Oct 2005

Ad-hoc-Arbeitsgruppe (AG) Boden (2005) Bodenkundliche Kartieranleitung. Bundesanstalt für Geowissenschaften und Rohstoffe in Zusammenarbeit mit den Staatlichen Geologischen Diensten, 5th ed. Hannover

Al-Ansari N, Shatnawi A (2011) Siltation of three small reservoirs in Jordan. Publications by the Institute of Earth and Environmental Sciences. Al-Bayt University Jordan, Jordan

Al-Ansari N, Alroubai A, Knutsson S (2012) Bathymetry and sediment survey for two old water harvesting schemes, Jordan. J Earth Sci Geotech Eng 2(2):13–23

Bochet E, García-Fayos P (2004) Factors controlling vegetation establishment and water erosion on motorway slopes in Valencia, Spain. Restor Ecol 12(2):166–174

Brune GM (1953) Trap efficiency of reservoirs, Eos Trans. AGU 34:407–418

Casalí J, López JJ, Giráldez JV (1999) Ephemeral gully erosion in southern Navarra, Spain. Catena 36:65–84

Cerdà A (1998) The influence of aspect and vegetation on seasonal changes in erosion under rainfall simulation on a clay soil in Spain. Can J Soil Sci, 321–330

Cerdà A (2002) The effect of season and parent material on water erosion on highly eroded soils in eastern Spain. J Arid Environ 52:319–337

De Rose R, Wilson DJ, Bartley R, et al (2005) Riverbank erosion and its importance to uncertainties in large scale sediment budgets. In: Walling D, Horowitz JA (eds) 2005. Sediment budgets 1, IAHS Publication 291. IAHS Press, pp 85–92

Ehiorobo JO, Audu HAP (2012) Monitoring of gully erosion in an urban area using geoinformation technology. J Emerg Trends Eng Appl Sci (JETEAS), 270–275

El-Radaideh NM (2010) Using bottom reservoir sediments as a source of agricultural soil: Wadi El-Arab reservoir as a case study, NW Jordan. Abhath Al-Yarmouk: Basic Sci Eng 19(2):75–91

Faust D, Schmidt M (2009) Soil erosion processes and sediment fluxes in a Mediterranean marl landscape, Campiña de Cádiz, SW Spain. Zeitschrift für Geomorphologie 52:247–265

Fentie B, Marsh N, Steven A, (2005) Sensitivity Analysis of a catchment scale sediment generation and transport model. In proceedings of: MODSIM 2005 international congress on modelling and simulation. Melbourne

Fryers K (2013) (Dis)Connectivity in catchment sediment cascades: a fresh look at the sediment delivery problem. Earth Surf Proc Land 38:30–46

García-Ruiz JM (2010) The effects of land uses on soil erosion in Spain: a review. Catena 81:1–11
García-Ruiz JM, Nadal-Romero E, Lana-Renault N, Beguería S (2013) Erosion in Mediterranean landscapes: changes and future challenges. Geomorphology 198:20–36
Ghrefat H, Yusuf N (2006) Assessing Mn, Fe, Cu, Zn, and Cd pollution in bottom sediments of Wadi Al-Arab Dam, Jordan. Chemosphere 65:2114–2121
Jenny H (1994) Factors of soil formation—a system of quantitative pedology. Dover Publication Inc., New York
Jetten V, Govers G, Hessel R (2003) Erosion models: quality of spatial predictions. Hydrol Process 17:887–900
Kosmas C, Danalatos N, Cammeraat LH, Chabart M, Diamantopoulos J, Farand R, Gutierrez L, Jacob A, Marques H, Martinez-Fernandez J, Mizara A, Moustakas N, Nicolau JM, Oliveros C, Pinna G, Puddu R, Puigdefabregas J, Roxo M, Simao A, Stamou G, Tomasi N, Usai D, Vacca A (1997) The effect of land use on runoff and soil erosion rates under Mediterranean conditions. Catena 29:45–49
Kralisch S, Krause P (2006) JAMS-A framework for natural resource model development and application. In: Proceedings of the iEMSs Third Biennial Meeting of "Summit on Environmental Modelling and Software", iEMSs, iEMSs, Manno, Switzerland
Kraushaar S, Herrmann N, Ollesch G, Vogel H-J, Siebert C (2014) Mound measurements—quantifying medium term soil erosion under olive trees in Northern Jordan. Geomorphology 213:1–12
McKergow L, Prosser I, Hughes A, Brodie J (2005) Sources of sediment to the Great Barrier Reef world heritage area. Mar Pollut Bull 51:200–211
Merritt WS, Letcher RA, Jakeman AJ (2003) A review of erosion and sediment transport models. Environ Model Softw 18:761–799
Meusburger K, Konz N, Schaub M, Alewell C (2010) Soil erosion modelled with USLE and PESERA using QuickBird derived vegetation parameters in an alpine catchment. Int J Appl Earth Obs Geoinf 12:208–215
Milne G (1936) A provisional Soil Map of East Africa (Kenya, Uganda, Tanganyika and Zanzibar) with explanatory memoir. Amani Memoirs, London
Ministry of Water and Irrigation (MWI) Jordan (2001) Options for artificial ground water recharge with reclaimed water in the Amman-Zarqa basin and Jordan valley. The Water Resource Policy Support, supported by USAID
Newham LTH, Norton JP, Prosser IP, Croke BFW, Jakeman AJ (2003) Sensitivity analysis for assessing the behavior of a landscape-based sediment source and transport model. Environ Model Softw 18:741–751
Nyssen J, Poesen J, Moeyersons J, Luyten E, Veyret-Picot M, Jozef D, Haile M, Grovers G (2002) Impact of road building on gully erosion risk: a case study from the northern Ethiopian Highlands. Earth Surf Proc Land 27:1267–1283
Obeidat SM, Alomary A, Sekhaneh W, Al-Momani I, Al Abdel Hamid AJ (2011) Assessment of water quality in four main water reservoirs in Northern Jordan. Int J Chem 3(2):79–87
Panagos P, Meusburger K, Alewellb C, Montanarella L (2012) Soil erodibility estimation using LUCAS point survey data of Europe. Environ Model Softw 30:143–145
Panagos P, Meusburger K, Ballabio C, Borrelli P, Alewell C (2014) Soil erodibility in Europe: a high-resolution dataset based on LUCAS. Sci Total Environ 479–480:189–200
Prosser IP, Rustomji P, Young B, Moran C, Hughes A (2001) Constructing river basin sediment budgets for the national land and water resource audit. CSIRO Land Water Technical Rep 15 (01):34
Renard KG, Foster GA, Weesies DK, et al (1997) Predicting soil erosion by water: a guide to conservation planning with the revised universal soil loss equation. Agriculture handbook 703. United States Department of Agriculture, Washington, DC
Renganayaki SP, Parimala S, Elango L (2013) A review on managed aquifer recharge by check dams: a case study near Chennai, India. Int J Res Eng Technol 2(4):416–423

Rustomji P, Caitcheon G, Hairsine P (2008) Combining a spatial model with geochemical tracers and river station data to construct a catchment sediment budget. Water Resour Res 44 (W01422):1–16

Rustomji P, Shellberg J, Brooks A et al (2010) A catchment sediment and nutrient budget for the Mitchell River, Queensland. Water for a healthy country report series ISSN: 1835–095X. CSIRO

Siebert C, Möller P, Rödiger T et al (2014) Thermal waters in the lower Yarmouk Gorge and their relation to surrounding aquifers. J Geochemis [unpublished]

Valentin C, Poesen J, Li Y (2005) Gully erosion: Impacts, factors and control. Catean 63:132–153

Vanmaercke M, Poesen J, Verstraeten G, de Vente J, Ocakoglu F (2011) Sediment yield in Europe: spatial patterns and scale dependency. Geomorphology 130:142–161

Wilkinson SN, Henderson A, Chen Y, Sherman B (2004) SedNet user guide, version 2 client report. CSIRO Land and Water, Canberra

Wilkinson SN, Young WJ, DeRose RC (2006) Regionalizing mean annual flow and daily flow variability for basin-scale sediment and nutrient modelling. Hydrol Process 20:2769–2786

Wilkinson S, Henderson A, Chen Y, Sherman B (2008) SedNet user guide. Client Report. CSIRO Land and Water, Canberra

Wilkinson SN, Prosser IA, Rustomji P, Read AM (2009) Modelling and testing spatially distributed sediment budgets to relate erosion processes to sediment yields. Environ Model Softw 24:489–501

Wischmeier WH, Smith DD (1978) Predicting rainfall erosion losses-a guide to conservation planning. Agriculture Handbook 537. United States Department of Agriculture

Wolf L, Werz H, Hoetzl H, et al (2007) Exploring the potential of managed aquifer recharge to mitigate water scarcity in the lower Jordan River basin within an IWRM approach. In: Fox P (ed) Management of aquifer recharge for sustainability. Proceedings of the 6th international symposium on managed artificial recharge of ground water, ISMAR6, Phoenix, Arizona/USA, Oct 28–Nov 2

Chapter 7
Final Conclusion and Discussion of the Results

The present study aimed to obtain estimates of erosion processes and sediment flux in the Mediterranean to semi-arid catchment Wadi Al-Arab in NW Jordan (263.5 km^2) and to evaluate the qualitative impact on the Wadi al-Arab reservoir. Measurements on the slope and in the sink, as well as a geochemical sediment fingerprint were compiled in a multi response approach to calibrate and validate the sediment transport model SedNet. The model allowed the spatial assessment of erosion hot spots and sediment yield, which helped to identify suitable areas for the construction of MAR structures in the Wadi bed.

The measurements were implemented in the major land use units (i) olive orchards (27 %), (ii) agricultural fields (14 %), and (iii) natural vegetated slopes under grazing (34 %). Olive orchards on steep slopes (>5–6°) showed soil loss between 72 ± 7 and 135 ± 19 t ha^{-1} $year^{-1}$, with an average value of 95 ± 8 t ha^{-1} $year^{-1}$. The cistern sediments date back to 760–862 AD, which is in agreement with archaeological findings on the abandonment of the region. The long-term average soil loss values range between 3.04 and 6.64 t ha^{-1} $year^{-1}$. The erosion pin plots showed an average of 23 ± 17.7 t ha^{-1} $year^{-1}$ of dislocated material, which partially re-sedimented within the plot boundaries. Stronger erosion was detected on south-facing slopes (Chap. 3).

In Chap. 4 the sink analysis showed that the Wadi Al-Arab reservoir has meromictic characteristics in summer with anaerobic conditions below a depth of 10 m. The average sedimentation rate was estimated to result in 67,000 m^3 ± 6 % per year, reducing the storage capacity by 9 % in the last 26 years. The predominantly clayey and silty sediments mainly accumulate close to the outlet of the reservoir at the dam wall in the west of the reservoir. This poses currently an acute threat to the controlled release of water to the agricultural areas in the Jordan Valley and reduces the life span of the reservoirs' function as there is no equipment for restoration (personal communication H. Al-Shurieki, dam director, 2012).

Qualitatively, sediments show a very high concentration of the heavy metals chromium, nickel, and cadmium, which are of geological origin. The high inorganic carbonatic background of the catchment strongly reduces the mobility of the heavy

S. Kraushaar, *Soil Erosion and Sediment Flux in Northern Jordan*,
Springer Theses, DOI 10.1007/978-3-319-31888-2_7

metals (Blume 2004). However, according to the European Policy for Effluent Sludge (2000), Blume (2004) the application of these sediments on agricultural used land would be prohibited, because plants can lower on small scale the pH and take up the heavy metals (personal communication Prof. Dr. Vetterlein, 2013).

Throughout the reservoirs' sediment cores the element concentrations stay constant in respect to their error spans. This proves that sediment sources have not changed in the past few years and are not yet exhausted.

Water samples confirm that common cations and anions show no noticeable concentrations and were in line with the quality standards of the German Drinking Water Directive (TrinkwV 2001) and the international WHO (2011) standards. However, a more detailed analysis of emerging compounds, such as cosmetic products (tonalide and galaxolide), pharmaceuticals (triclosan and carbamazepine), as well as herbicides and insecticides (hexazinone, t–NP, trifluralin), confirm the contamination of the water by these anthropogenic pollutants. Cosmetics and pharmaceuticals enter the reservoir through observed waste water effluents from the catchments' two treatment plants, illegal sewage water spill, and damaged sewage tanks. The chemicals used in agriculture are sediment bound and derive from diffusive sediment sources throughout the catchment.

The sediment fingerprint method (Chap. 5) allowed geochemically differentiating 6 sediment sources and calculating their relative contribution to the sink. The method relies on the comparability of the sediments on the slope with those in the sink. However, processes like grain size selectivity, solubility, or chemical interaction prohibit direct comparison and a correction factor approach was developed. Using a "stepwise multiple regression analysis model" (SMRAM), parameters such as clay and inorganic and organic carbonate content were identified to be altered during transport. Their volume change influences the chemical concentrations of the sink samples and hence needed to be numerically corrected. Additionally, a course of action was suggested for the detection of non-soluble and conservative elements that differ significantly in concentrations between the sources. Applying this decision tree, 7 out of a possible 28 elements allowed the significant characterization of the 6 sources. These are: Si, Mn, Fe, Ni, ^{232}Th, ^{228}Th, ^{137}Cs. They differ in concentration between the sources due to their geological background, nature of input (e.g. natural airborne radio nuclides, anthropogenic input etc.), land use and related agricultural treatment (e.g. homogenization of concentrations in the topsoil through ploughing), and soil development (enrichment of conservative elements as Fe or Al in geomorphological stable positions due to soil development).

Results of the mixing model show that mainly shallow yellow Mediterranean soils as they exist on steep vegetated slopes used for grazing or olive tree plantation contribute to the reservoir sediments with around 70 ± 17 %. There of, 59 ± 8 % are delivered from land under olive trees and 29 ± 15 % from slopes under grazing. Agricultural used plateaus with red Mediterranean soils contribute to the sink with 11 ± 9 %. With regard to their error spans, no sediments originated from unconsolidated geological sources following the model result. Reasons could be that samples are not representative of the chosen sources, or subsoil horizons and initial sediment are not major contributors because gullying is not the predominant erosion process in the region. The model proved to be quite robust to uncertainties with a mean residual error of 1.04 %.

The implementation of the fingerprint revealed that ^{210}Pb, which is believed to be conservative and hence often used for sediment fingerprint studies is non conservative in the Wadi Al-Arab. Due to the high uranium present in the phosphorites of the carbonatic geology, this element (^{238}U) and its daughter nuclide, ^{210}Pb, are added to the sink in an unknown amount and thus proven to be non-conservative and not useful for this study.

Reservoir sediments show ^{137}Cs concentrations comparable to those of the topsoil samples (17.2 ± 1.0 Bk kg^{-1}). Temporal sediment storages as the terraces in the last km of the Wadi bed show no ^{137}Cs content but they do show red soils (7.5 YR 4/4), indicating geomorphological stable positions and soil development. From these results it is possible to conclude that the reservoir sediments lay once exposed on the surface in the last 60 years. Contrastingly, the intermediate storages such as terraces have not been eroded in the last 60 years. Otherwise one would have expected them to have diluted the ^{137}Cs concentrations in the sink reservoir. This indicates a high connectivity of the slopes to the river bed and, hence, transport of material from the slopes to the sink. Additionally, it highlights the high sensitivity and vulnerability of the system to changes on the slope that trigger increased erosion, such as, for example, olive plantings on steep slopes under ploughing.

Finally, and following the question of how erosion and sediment yield is spatially developed throughout the catchment, the Sediment River Network Model (SedNet) was implemented for the Wadi Al-Arab catchment (Chap. 6). Detailed soil and gully mapping allowed the catchment-wide estimation of sensitive model parameters such as sediment bulk density, certain grain size fractions, as well as gully morphology. The sediment delivery ratio was modelled using satellite data reflecting vegetation cover and declination. Calibration and validation was done with field findings using the average yearly sedimentation rate in the Wadi Al-Arab reservoir, and the model results were compared to the erosion measurements from the major land use units. Results show that the model is capable of predicting erosion on the slope quite well. The reason for the mean differences in the olive orchard erosion values lies in the nature of the mound measurement method used for the calculation of average medium-term erosion rates. The method is only suitable for steep slopes >5–6° (Sect.3.2) and therefore correspond correctly only to the highest modelled erosion rates.

Total mean erosion rates for the sub-catchments evaluated with the SedNet model range between 1.2 and 8.5 t ha^{-1} year^{-1}. These values correspond to the realm of possibilities of soil erosion rates measured on different land use units in the Mediterranean (Garcìa-Ruiz et al. 2013). Furthermore, the averages fall below the tolerable soil loss as defined by the U.S. Department of Agriculture (=5–12 t ha^{-1} year^{-1} T value; Montgomery 2007) and by Moreira-Madueño for different soils in Spain (=8.4 and 11.4 t ha^{-1} year^{-1}; Gómez et al. 2003). However, as discussed in Sect. 3.2, many scholars would estimate tolerable soil loss in a much lower amount (=0.3–2. 6 t ha^{-1} year^{-1}; Jürgens and Fander 1993; Visser et al. 2004; Montgomery 2007; Verheijen et al. 2009). In regard to these contrasting thresholds and the variety of possible erosion rates, the definition of sediment sources and the pattern of erosion affected areas are of greater importance. This knowledge, together with the estimation of sediment connectivity in each sub-catchment, enabled delineating areas sensitive to land use changes, in regard to the total sediment yield in the riverbed.

The patterns show that especially the north, northwest, and west of the catchment suffer from high erosion values from slopes and gullies, of which hillslopes serve as the major contributor. Erosion results correspond strongly to the extension of the marly geological unit MCM, increase in relief energy, and the extent of the land use units olive orchards and steep slopes.

Areas most suitable for the installation of in channel modifications, such as check dams to slow down runoff, reduce erosion, and increase infiltration into lower aquifers, were identified. Criteria included (i) the low soil erosion on the hillslopes, (ii) the low sediment connectivity of the hillslopes to the river, and hence low sediment yield, (iii) a headwater position not downstream of a Waste Water treatment plant, (iv) nor above a spring. Sub-catchments with head waters in the east and middle of the catchment suit these conditions best.

References

Blume HP (Ed) (2004) Handbuch des Bodenschutzes. Bodenökologie und–belastung—Vorbeugende und abwehrende Schutzmaßnahmen, 3rd edn. Ecomed, Landsberg am Lech, 723–769

García-Ruiz JM, Nadal-Romero E, Lana-Renault N et al. (2013) Erosion in Mediterranean landscapes: changes and future challenges. Geomorphology 198:20–36

Gómez JA, Battany B, Renschler CS, Fereres E (2003) Evaluating the impact of soil management on soil loss in olive orchards. Soil Use Manag 19:127–134

Jürgens C, Fander M (1993) Soil erosion assessment by means of LANDSAT-TM and ancillary digital data in relation to water quality. Soil Technology 6:215–223

Montgomery DR (2007) Soil erosion and agricultural sustainability. Proceeding of the National Academy of Science of the United States of America 104(33):13268–13272

Trinkwasserverordnung (TrinkwV) (2001) http://www.gesetze-im-internet.de/trinkwv_2001/index.html (last checked 05.05.2014)

Verheijen FGA, Jones RJA, Rickson RJ, Smith CJ (2009) Tolerable versus actual soil erosion rates in Europe. Earth Sci Rev 94:23–38

Visser SM, Sterk G, Ribolzi O (2004) Techniques for simultaneous quantification of wind and water erosion in semi-arid regions. J Arid Environ 59:699–717

World Health Organization (WHO) (2011) Guidelines for Drinking-Water Quality, 4th edn. WHO Press, Geneva, Switzerland

Chapter 8
Outlook

Especially the intensification of olive plantations in the last 25 years on less suitable land with steep slopes (>5°) and marly substrate led to stronger erosion of this land use unit in the region. This could be confirmed by the measurements in the olive plantations (Sect. 3.2), the sediment fingerprint (Chap. 5), and the SedNet model (Chap. 6). The simulation in SedNet of a land use scenario with adapted olive farming showed that this management change would cause a 54 % reduction (60,000 t year^{-1}) of the overall annual sediment load in the dam. This high impact is explained with the concentration of especially young and highly erosive olive orchards in the north and west of the catchment. The areas coincide with the high sediment connectivity of these respective sub-catchments and thus are particularly sensitive to changes of land use and management. Hence, the single management alteration in olive plantations (27 % of the catchment) causes such a strong impact on the total sediment yield.

The results stress the fact that adapted farming strategies in olive orchards and other areas of increased erosion, such as slopes used for grazing, are of special relevance to the region. This becomes even more important in regard to the following postulated future developments:

1. Market liberalisation: A trend for further intensification of olive plantations is expected with the expansion of the European trading zone to include Jordan in 2014 (Jacquet et al. 2007). This is supported by the fact that in the 1990s, Jordan introduced a governmental policy to lower the price of olive seedlings in order to "stem rural depopulation by providing jobs for roughly 6 % of the working population" (Venot 2003; IOC 2005; Sect. 3.2).
2. Population increase: In the light of strong population increase (around 9 % since 2012, UNHCR 2014) due to refugee waves coming in from Syria, effects on agriculture such as higher meat demand and thus grazing activities, as well as the increased availability of cheap labourers for the intensification of agriculture, are imaginable.

S. Kraushaar, *Soil Erosion and Sediment Flux in Northern Jordan*,
Springer Theses, DOI 10.1007/978-3-319-31888-2_8

3. Aridisation: Langbein and Schumm (1958) as well as Kosmas et al. (1997) proved in their comprehensive studies that runoff and sediment yields reach their maximum when average rainfall ranges between 200 and 400 mm year^{-1}. This amount of rain is enough to generate runoff but is too little to maintain a protective vegetation cover in most areas. Currently the Wadi Al-Arab receives 390–495 mm of rain. However, results by Kunstmann (2007) and Hörling et al. (2012) indicate further aridisation of the region, which in return implies an even higher sediment yield from slopes in future years.

However, adapted management will only find its way into practice if legislation and executive powers find ways to motivate farmers. Abu-Sharar (2006) identified several deficiencies concerning the primary environmental Law of Jordan No. 12 (1995) and the temporary Environment Law No. 1 (2003). They include the missing articles dealing with soil conservation and desertification, absence of articles dealing with crimes of environmental nature like soil contamination, civil responsibilities, and compensation for environmental damages as well as the nonexistence of a comprehensive legislation concerning land use. Furthermore, he states that the interdepartmental and multiple responsibilities, which exist within the environmental sector, together with the lack of coordination, hinder enacting laws, bylaws, and instructions. Ambiguities in some law articles help to avoid penalties. Additionally, slow court proceedings and minor penalties compared to the magnitude of the environmental damage they deal with, result in more law violations (Abu-Sharar 2006). Public participation in environmental policy-making, a review of the environmental legislation, and the establishment of an environmental fund to combat soil degradation are just a few ideas to tackle the problem and encourage the right motivation for the public and the industry to care about soil sustainability and adapt management strategies (Abu-Sharar 2006). Additionally, an executive authority needs to be present in adequate numbers to follow up on the adherence of the laws (personal communication Dr. Lee, NICE office Amman, 2014).

Last but not least, Abu-Sharar (2006) emphasize that scientific references for argued environmental cases are necessary. In this respect, much still has to be done in Jordan in the sector of soil erosion and land degradation and future research should focus on the effects of recent land use changes on sediment yield as well as the relevant factors of erosion. For management decisions on a broad scale, modelling will remain a relevant tool demanding the incorporation of site specific Mediterranean landscape and soil characteristics such as sodium content into the K-factor of the RUSLE model. Colluvial sediment archives, such as debris cones on footslope positions, which are present in the western part of the catchment around the reservoir, could give further answers to the development of erosion throughout the centuries. This could be an additional focus in future studies that aim to tackle erosion on the slope scale.

References

Abu-Sharar TM (2006) The challenges of land and water resources degradation in Jordan: diagnosis and solutions. In: Kepner WG, Rubio JL, Mouat DA, Pedrazzini F (eds) Desertification in the Mediterranean. Springer, Netherlands, pp 201–226

Hörling M, Eischeid J, Perlwitz J, Quan X (2012) On the increased frequency of Mediterranean droughts. J Clim 25:2146–2160

IOC (International Oil Council) (2005) Country profiles—Jordan E.108/Doc. no. 4, Update no. 31, Ref.: 18,1–17. http://www.internationaloliveoil.org/ (last checked 29.08.2011)

Kunstmann H (2007) Regionale Auswirkungen der Klimaänderung auf die Wasserverfügbarkeit in Klimasensitiven Gebieten. In: Endlicher W, Gerstengarbe FW (eds) Der Klimawandel—Einblicke, Rückblicke und Ausblicke. Institut für Klimafolgenforschung, Potsdam

Langbein WB, Schumm SA (1958) Yield of sediment in relation to mean annual precipitation. Am Geophys Union 39(6):1076–1084

The UN Refugee Agency (UNHCR) (2014) http://data.unhcr.org/syrianrefugees/country.php?id=107 (last checked 10.06.2014)

Venot JP (2003) The peculiar case of the olive trees orchard. Reclamation's history of the Jordan river basin in Jordan, a focus on agriculture: past trends, actual farming systems and future prospective. Vol IV. INA P-G Paris-Grignon National Institute of Agronomy, French Regional Mission for Water and Agriculture (MREA) French Embassy, International Water Management Institute (IWMI)

Chapter 9
Innovative Aspects of the Research

The findings of the measurements qualitatively correspond to the geochemical sediment fingerprint and are reflected in the model outcomes. Results confirmed and complemented each other and thus the method compilation guaranteed a more holistic understanding of the sediment fluxes in the Wadi Al-Arab catchment. This allowed the observation of erosion and sedimentation from different methodological perspectives. Furthermore, the research resulted in the following new methodological aspects:

- The analysis and dating of cistern sediments in regard to erosion rates has not been done before. The abundance of the Roman cisterns as mapped for this study and documented for the Mediterranean region constitute a great advantage when aiming at numerous long-term average soil erosion rates of nearly levelled positions and across the region.
- The proof of concept of the olive mound method was done in a region other than the one chosen by Vanwalleghem (2010), who implemented the method under olive trees in southern Andalusia. Concrete criteria are formulated for the detection of olive mounds that are residual erosion markers and not anthropogenic accumulation.
- The aforementioned methods cover different time scales: 560 years for the fields, and an average of 32 years for the olive orchards. These long time spans have considerable advantages compared to many conventional erosion measurements, such as bounded plot and sediment trap installations. They are non- to minimally invasive, and, hence, no boundary or edge effects occur. Furthermore they include all possible erosion processes and deliver long-term average erosion estimates that do not underlie short-term weather fluctuations as they can occur in the course of a three-year time period of a Ph.D. Thesis.
- Concerning the sediment fingerprint, the newly suggested method (SMRAM) allows the statistical justified identification and correction of influencing parameters, while taking possible interrelations between them into account. Further developments include the conservative assessment of the elements in

S. Kraushaar, *Soil Erosion and Sediment Flux in Northern Jordan*,
Springer Theses, DOI 10.1007/978-3-319-31888-2_9

comparison to the water chemistry and their bonding tendency to other soluble elements. The revised approach is of special relevance to areas where, due to data scarcity, little is known about chemical sediment characteristics and interdependencies.

– The construction of a parameterized and digital soil map with aspect dependent soil information and the consequent inclusion of the aspect dependence in the C-factor of the RUSLE have not been implemented before. Furthermore, the SedNet model has been parameterized for this region for the first time.

In summary, the results of this work are one of the first integrated studies on recent soil erosion in an exemplary catchment in Jordan. The study alerts authorities of certain agricultural subsidies and market developments that might result in land use changes and practices. It includes important information for policy-makers and resource authorities when planning actions to deal with soil loss and increasing sedimentation in reservoirs. Finally, it alludes to the importance of soil erosion and sediment yield concerning the pollution of surface water bodies as well as MAR strategies focusing on in channel constructions.

References

Climate-Data. http://en.climate-data.org/location/3388/ (last checked 28.05.2014)

Döring M (2009) Der längste Tunnel der antiken Welt. Antike Welt 2:26–34

IV-G. http://iv-g.livejournal.com/295425.html (last checked 28.05.2014)

IWRM-SMART2. http://www.iwrm-smart2.org/index.php/outline/working-area (last checked 15. 01.12)

Merritt WS, Letcher RA, Jakeman AJ (2003) A review of erosion and sediment transport models. Environ Model Softw 18:761–799

Rödiger T, Geyer S, Mallast U, Merz R, Krause P, Fischer C, Siebert C (2014) Multi-response calibration of a conceptual hydrological model in the semiarid catchment of Wadi al Arab, Jordan. J Hydrol 509:193–206

Vanwalleghem T, Laguna A, Giráldez JV, Jiménez-Hornero FJ (2010) Applying a simple methodology to assess historical soil erosion in olive orchards. Geomorphology 114:294–302

Working Group on water issues of the Federal States and the Federal Government (LAWA) (1992) Grundwasserrichtlinie, Teil 3: Grundwasserbeschaffenheit

Appendix Chapter 2

Wadi Al-Arab Geology (Moh'd 2000)

Geological unit	Age	Period	Stratigraphic nomenclature	Hydrological unit	Characteristics (Moh'd 2000)
Calcrete/ Travertine/ Soils	Recent	Quaternary		Travertine as indicator for aquifer drainage	Soils (FAO 1979): Vertisols, calcaric and vertic Cambisols, eutric Lithosols, haplic and calcaric Phaeozems
Basalt	Pliocene—recent	Quarternary		Local Aquifer	Plateau Basalts as cover basalts conserve the existing relief, no basalt columns
Waqqas Conglomerate	Oligocene	Paleogene	Jordan Valley Group	Local Aquifer	Ca. 200 m thick 1. (calcerous) sandstone 2. marl with gravel 3. gravels and conglomerates with limestone elements
Umm Rijam Chert Limestone	Paleocene/Eocene	Paleogene	Balqa Group	B4 = Aquifer	Ca. 220 m thick 1. marly chalk 2. chalky limestone 3. upper chert unit, bitumen present

(continued)

S. Kraushaar, *Soil Erosion and Sediment Flux in Northern Jordan*,
Springer Theses, DOI 10.1007/978-3-319-31888-2

(continued)

Geological unit	Age	Period	Stratigraphic nomenclature	Hydrological unit	Characteristics (Moh'd 2000)
Muwaqqar Chalk Marl	Maastrichtian	Upper Cretaceous	Balqa Group	B3 = Aquitard	Ca. 100 m thick 1. non bituminous chalk marl 2. bituminous chalk marl 3. nodular-bedded limestone 4. chalky limestone cliff
Amman Silicified Limestone/Al Hisa Phosphorites	Santonian/Campanian	Upper Cretaceous	Balqa Group	B2 = Aquifer	ASL: ca. 60 m thick Alternating beds of silicified chert and limestone beds, highly fractured AHP: ca. 10 m thick 1. bedded phosphorites 2. phosphorites alternating with marl 3. thick chalk High in Uranium content
Wadi Umm Ghudran	Turonian	Upper Cretaceous	Belqa Group	B1 = Aquifer	Ca. 40m thick Fossoliferous 1. chalk 2. chalk beds with coquinal limestone and dolomitic chalky limestone
Wadi As Sir Limestone	Turonian	Upper Cretaceous	Ajloun Group	A7 = Aquifer	>200 m thick, strongly carstified 1. dolomitized limestone 2. limestone 3. peloidal limestone (sometimes marly or with cherts) 4. micritic limestone

Moh'd BK (2000) The geology of Irbid and Ash Shuna Ash Shamaliyya (WAQQAS), Map Sheets No. 3154-II and 3154-III. Geological Directorate. Geological Mapping Division, Bulletin 46, Amman

Food and Agriculture Organization of the UN (FAO) (1979) Soil classification http://www.fao.org/soils-portal/soil-survey/soil-classification/en/ (last checked 12. 06.2014)

Appendix Chapter 3

Section 3.1

S. Kraushaar, *Soil Erosion and Sediment Flux in Northern Jordan*,
Springer Theses, DOI 10.1007/978-3-319-31888-2

Radio nuclide content of soil samples from the Roman ***Cistern 1***

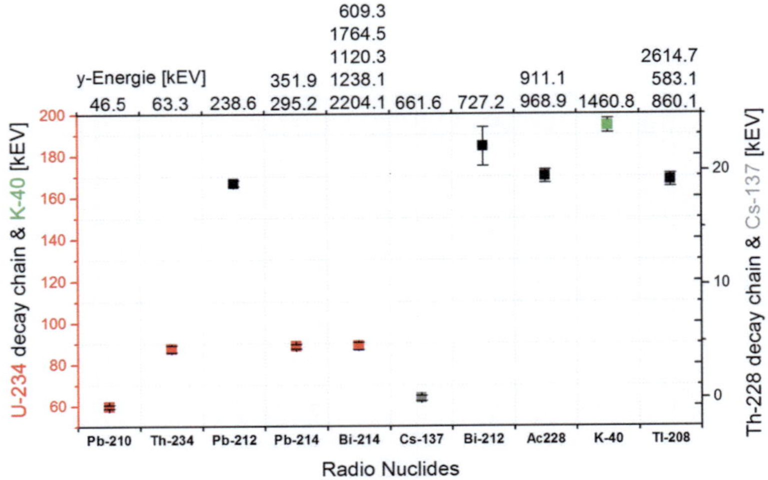

Radio nuclide content of soil samples from the Roman ***Cistern 2***

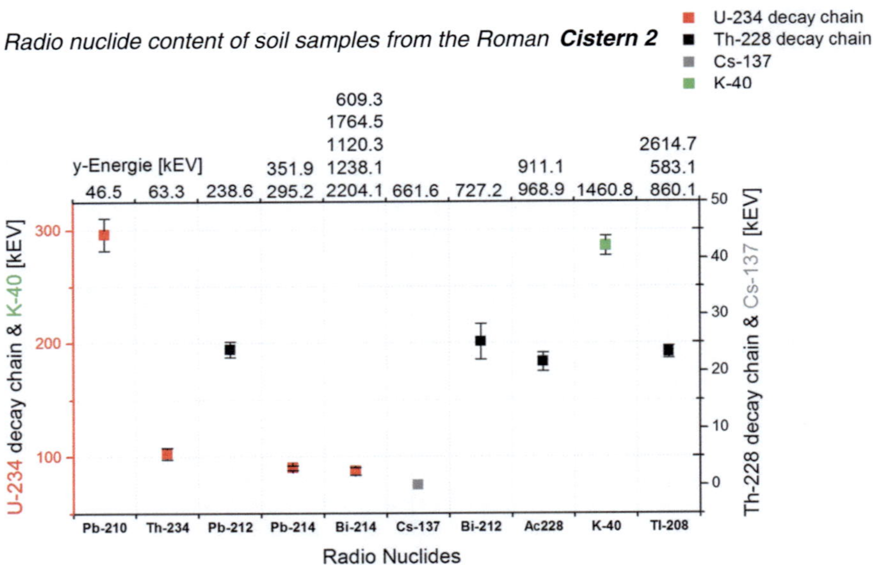

Section 3.2

Soil samples from the olive mound fields

Field	Location	Clay (%)	Silt (%)	Sand (%)	Grain size[a]
Tell Zira'a	Intra-mound	14.64	77.73	7.64	Silt loam
Tell Zira'a	Intra-mound	14.40	77.22	8.39	Silt loam
Tell Zira'a	Intra-mound	14.58	75.18	10.25	Silt loam
Tell Zira'a	Intra-mound	13.94	76.35	9.72	Silt loam
Tell Zira'a	Intra-mound	13.29	74.76	11.95	Silt loam
Tell Zira'a	Intra-mound	12.69	79.34	7.98	Silt loam
Isara	Intra-mound	10.59	84.31	5.11	Silt
Isara	Mound	6.04	88.96	5.01	Silt
Isara	Intra-mound	10.13	85.85	4.03	Silt
Isara	Intra-mound	10.11	85.24	4.66	Silt
Isara	Mound	10.34	79.39	10.27	Silt
Rusan	Mound	11.81	80.97	7.22	Silt loam
Rusan	Intra-mound	15.73	71.31	12.96	Silt loam
Rusan	Mound	14.32	70.97	14.72	Silt loam
Rusan	Intra-mound	15.49	71.00	13.52	Silt loam
Rusan	Mound	12.30	71.98	15.73	Silt loam
Rusan	Intra-mound	14.71	77.70	7.60	Silt loam
Malka	Mound	12.66	73.11	14.24	Silt loam
Malka	Intra-mound	16.24	77.61	6.15	Silt loam
Malka	Mound	13.81	74.94	11.25	Silt loam
Malka	Intra-mound	16.40	78.55	5.06	Silt loam
Sifin I	Intra-mound	20.64	76.20	3.17	Silt loam
Sifin I	Mound	18.43	69.29	12.29	Silt loam
Sifin I	Intra-mound	15.80	78.27	5.94	Silt loam
Sifin I	Mound	16.39	72.54	11.07	Silt loam
Sifin II	Mound	8.69	61.55	29.77	Silt loam
Sifin II	Intra-mound	19.79	75.36	4.86	Silt loam
Sifin II	Mound	17.23	73.09	9.69	Silt loam
Sifin II	Intra-mound	18.86	74.51	6.64	Silt loam

[a]Grain size classes according to FAO (1979)

Soil samples from the olive mound fields

Field	Location	Bulk density (g cm^{-3})	Kf[a] (m d^{-1})	Field capacity[b] (g)	Field capacity[b] (%)
Tell Zira'a	Intra Mound	1.03	0.21	48.14	41.77
Tell Zira'a	Intra Mound	1.12	0.09	46.47	35.23
Tell Zira'a	Intra Mound	1.13	0.06	51.48	42.29
Tell Zira'a	Intra Mound	1.25	0.04	52.27	38.45
Tell Zira'a	Intra Mound	1.31	0.05	53.84	37.12
Tell Zira'a	Intra Mound	1.19	0.06	51.33	39.09
Tell Zira'a	Intra Mound	1.16	0.12	51.17	39.00
Tell Zira'a	Intra Mound	1.20	0.10	52.71	40.42
Tell Zira'a	Intra Mound	1.24	0.07	52.18	38.51
Tell Zira'a	Intra Mound	1.23	8.48	49.92	36.29
Tell Zira'a	Intra Mound	1.16	0.08	50.05	37.34
Tell Zira'a	Intra Mound	1.20	0.07	53.99	41.99
Tell Zira'a	Intra Mound	1.27	0.07	53.00	38.86
Isara	Intra Mound	1.16	0.18	68.83	57.50
Isara	Intra Mound	1.19	0.24	70.07	56.56
Isara	Intra Mound	1.03	1.02	67.02	63.04
Isara	Intra Mound	1.09	0.36	68.99	62.18
Isara	Mound	1.19	0.10	61.33	49.01
Isara	Intra Mound	1.18	0.07	66.63	54.97
Isara	Intra Mound	1.36	0.05	63.61	44.40
Isara	Mound	1.23	0.10	65.45	50.82
Isara	Intra Mound	1.12	0.10	56.71	44.78
Isara	Mound	0.99	0.12	56.95	53.54
Isara	Intra Mound	1.36	0.02	59.38	41.96
Isara	Mound	1.02	1.15	48.38	40.66
Rusan	Mound	1.11	17.81	53.51	44.00
Rusan	Intra Mound	1.22	0.04	58.13	44.70
Rusan	Mound	1.11	0.16	55.80	44.60
Rusan	Intra Mound	1.16	0.07	53.25	40.63
Rusan	Mound	1.10	0.13	49.73	41.44
Rusan	Intra Mound	1.23	0.14	56.04	41.93
Rusan	Mound	1.04	0.31	53.79	48.77
Rusan	Intra Mound	1.11	0.04	51.35	39.87
Rusan	Mound	1.12	0.48	49.37	40.68
Rusan	Intra Mound	1.09	0.15	55.32	47.59
Rusan	Mound	1.12	0.26	55.01	42.89
Rusan	Intra Mound	1.18	0.09	56.28	43.42

[a]Temperature corrected for 25 °C
[b]at pF 1.8

Soil samples from the olive mound fields

Field	Location	Bulk density (g cm^{-3})
Malka	Intra Mound	1.03
Malka	Mound	0.82
Malka	Intra Mound	0.92
Malka	Mound	0.92
Malka	Intra Mound	0.76
Malka	Mound	1.03
Malka	Mound	0.93
Malka	Intra Mound	0.97
Malka	Mound	0.89
Malka	Intra Mound	0.83
Malka	Mound	0.89
Malka	Intra Mound	0.90
Sum	Mound	1.13
Sum	Intra Mound	1.15
Sum	Mound	1.11
Sum	Intra Mound	0.99
Sum	Mound	1.09
Sum	Intra Mound	1.17
Sum	Mound	1.25
Sum	Intra Mound	1.14
Sum	Mound	1.19
Sum	Intra Mound	1.18
Sum	Intra Mound	1.14
Sum	Mound	1.23
Sifin I	Intra Mound	1.12
Sifin I	Mound	0.88
Sifin I	Intra Mound	0.91
Sifin I	Mound	0.82
Sifin I	Intra Mound	1.00
Sifin I	Mound	1.00
Sifin I	Intra Mound	1.02
Sifin I	Mound	1.03
Sifin I	Mound	1.10
Sifin I	Intra Mound	0.92
Sifin II	Mound	0.79
Sifin II	Intra Mound	0.93
Sifin II	Mound	0.86
Sifin II	Intra Mound	1.15
Sifin II	Mound	1.08

(continued)

(continued)

Field	Location	Bulk density (g cm^{-3})
Sifin II	Intra Mound	1.26
Sifin II	Mound	1.10
Sifin II	Intra Mound	1.20
Sifin II	Mound	0.93
Sifin II	Intra Mound	1.26

Food and Agriculture Organization of the UN (FAO) (1979) Soil classification http://www.fao.org/soils-portal/soil-survey/soil-classification/en/ (last checked 12. 06.2014)

Appendix Chapters 4 and 5

Soil, geology and sediment samples from the Wadi Al-Arab catchment

Samples		(mg kg^{-1})							
		LOI	Na	Mg	Al	Si	P	S	K
Surface near soil/sediment samples	Olive_01	326,300	710	9620	22,350	108,790	3560	825	3120
	Olive_02	351,900	980	8500	21,650	80,340	4500	601	4430
	Olive_03	322,100	620	9670	30,720	100,980	3310	460	4670
	Olive_04	338,100	910	5870	27,740	94,960	1720	345	4130
	Olive_05	376,300	530	4490	13,480	69,240	1890	645	2170
	Olive_06	201,700	1490	7450	66,330	203,450	2290	384	6620
	Field_01	229,900	1390	12,110	50,360	175,270	1030	<300	6130
	Field_02	302,400	960	6340	36,500	109,160	4620	351	3400
	Field_03	237,200	1460	7070	52,930	168,630	2730	419	7090
	Field_04	214,900	1410	7530	55,960	180,150	3230	<300	6280
	Field_05	310,500	1030	6310	33,360	110,440	6440	436	4680
	Field_06	126,200	1650	6650	72,480	272,140	1280	<300	5840
	Slope_01	379,800	1080	5400	25,430	72,690	5250	616	3250
	Slope_02	371,600	810	3640	24,240	63,780	4800	541	2000
	Slope_03	362,900	940	3770	29,290	79,650	9640	602	2030
	Slope_04	375,500	650	6760	19,660	66,080	4770	436	2220
	Slope_05	320,200	960	4390	35,980	98,900	6600	471	3320
	Slope_06	349,000	1100	5470	29,670	106,290	1910	701	3300
Geology samples	UR_01	429,600	2200	6400	2960	11,310	2130	<300	350
	UR_02	449,600	310	2290	2090	4630	1240	572	190
	UR_03	443,600	300	3160	2860	10,130	400	672	160
	UR_04	446,100	310	2940	2480	8620	380	753	240
	UR_05	442,400	340	2150	2040	6690	620	696	130
	UR_06	449,300	420	2230	1960	5750	520	447	150
	MCM_01	433,600	290	5530	3430	8860	2330	719	220
	MCM_02	430,700	9810	2840	9890	20,200	1890	881	790
	MCM_03	395,100	750	2240	12,370	27,470	7250	1427	1200
	MCM_04	393,600	700	3050	8630	39,880	1970	265	1220
	MCM_05	408,700	630	1700	10,350	23,460	1700	<300	1090

(continued)

S. Kraushaar, *Soil Erosion and Sediment Flux in Northern Jordan*,
Springer Theses, DOI 10.1007/978-3-319-31888-2

(continued)

Samples		(mg kg^{-1})							
		LOI	Na	Mg	Al	Si	P	S	K
	MCM_06	429,600	300	1450	6610	13,510	1410	<300	650
	ASL_01	460,200	650	5490	10,140	89,270	5510	23,403	5030
	ASL_02	245,200	500	1130	3150	174,870	12,060	<300	260
	ASL_03	426,600	420	1980	2600	12,260	4840	374	210
	ASL_04	426,900	3290	2320	4790	15,650	3200	354	240
	ASL_05	283,100	380	1130	2310	151,250	6450	<300	240
	ASL_06	356,200	660	1600	5020	57,470	12,110	720	540
Wadi bed	WD_01	404,400	530	3040	9930	39,430	2770	390	1590
	WD_02	319,200	790	4180	13,660	97,920	10,830	615	1670
	WD_03	330,800	1080	6050	22,200	91,460	6820	729	3640
						< below detection limit			

WD = Wadi Deposits

Soil, geology and sediment samples from the Wadi Al-Arab catchment

Samples	(mg kg^{-1})									
	Ca	Ti	V	Cr	Mn	Fe	Co	Ni	Cu	Zn
Olive_01	230,550	1440	67	116	838	16,160	<10	130	35	130
Olive_02	260,110	2070	42	78	360	15,460	<10	199	27	437
Olive_03	234,170	1700	77	143	371	19,160	<10	87	32	100
Olive_04	242,860	2780	58	92	408	21,120	<10	50	31	103
Olive_05	285,490	870	26	70	887	9650	<10	86	26	64
Olive_06	86,690	6640	146	197	892	50,780	19	80	42	139
Field_01	137,050	5860	116	127	731	40,570	12	71	37	109
Field_02	220,510	3430	121	205	411	27,700	<10	85	48	161
Field_03	142,790	5330	109	167	754	40,080	<10	71	38	137
Field_04	134,600	5710	139	205	766	42,250	14	78	41	168
Field_05	221,680	3000	110	170	411	23,850	<10	114	47	328
Field_06	24,600	8440	187	164	1209	54,770	22	85	40	205
Slope_01	250,480	1950	92	217	293	12,940	<10	81	49	171
Slope_02	269,980	1730	86	220	172	16,370	<10	102	55	196
Slope_03	245,510	1850	105	303	149	18,610	<10	131	79	271
Slope_04	273,260	1580	91	157	209	14,760	<10	55	35	122
Slope_05	226,240	2710	168	362	307	26,580	<10	124	80	223
Slope_06	218,760	2620	135	156	297	19,930	<10	77	28	271
UR_01	369,830	220	<10	70	33	1540	<10	14	18	58
UR_02	380,440	160	<10	35	16	700	<10	4	18	36
UR_03	373,290	210	14	18	24	1330	<10	17	18	45
UR_04	373,570	170	<10	4	96	840	<10	<2.5	10	16

(continued)

(continued)

Samples	(mg kg^{-1})									
	Ca	Ti	V	Cr	Mn	Fe	Co	Ni	Cu	Zn
UR_05	378,740	170	10	16	17	420	<10	6	15	33
UR_06	380,570	150	<10	16	16	280	<10	<2.6	16	25
MCM_01	369,650	220	<10	34	25	1540	<10	5	12	30
MCM_02	340,610	490	77	141	26	8040	<10	50	43	76
MCM_03	343,580	680	107	338	40	7900	<10	91	89	266
MCM_04	340,330	530	40	37	1007	6020	<10	67	23	68
MCM_05	354,920	470	99	149	40	10,350	<10	62	48	103
MCM_06	368,940	340	49	92	24	5460	<10	24	29	47
ASL_01	210,890	1280	127	257	38	5320	<10	148	58	736
ASL_02	233,790	250	72	126	17	1680	<10	302	19	150
ASL_03	376,060	200	58	43	14	700	<10	3	35	74
ASL_04	366,780	310	297	142	8	1540	<10	9	36	130
ASL_05	254,080	240	39	154	9	560	<10	346	23	174
ASL_06	314,850	320	81	130	8	2170	<10	27	27	151
WD_01	335,870	990	32	90	205	7340	<10	39	23	86
WD_02	263,470	1390	55	118	202	12,450	<10	51	38	180
WD_03	249,750	2250	106	160	352	17,420	<10	79	47	208
						< below detection limit				

WD = Wadi Deposits

Soil, geology and sediment samples from the Wadi Al-Arab catchment

Samples	(mg kg^{-1})													
	Se	As	Br	Rb	Sr	Zr	Pb	U	Th	Mo	Cd	Hg	Sn	Sb	Ba
Olive_01	<2	3	6	19	641	64	10	4	4	<2	4	<2	9	12	5738
Olive_02	2	4	12	28	407	93	12	3	4	<2	3	<2	3	<2	354
Olive_03	<2	<2	8	27	744	70	10	<2	5	<2	<2	<2	6	3	3274
Olive_04	<2	3	8	26	260	135	9	<2	5	<2	<2	<2	<2	<2	283
Olive_05	<2	<2	8	11	551	34	6	3	<2	<2	2	<2	5	5	4785
Olive_06	<2	11	8	52	130	318	15	3	8	2	<2	<2	4	<2	413
Field_01	<2	5	9	41	247	236	12	<2	7	<2	<2	<2	3	<2	504
Field_02	2	9	10	26	334	146	8	6	5	<2	<2	<2	3	<2	328
Field_03	2	6	9	48	183	251	15	<2	7	3	<2	<2	4	<2	498
Field_04	<2	7	9	45	262	270	13	5	8	10	5	<2	3	<2	375
Field_05	<2	10	9	34	259	141	11	7	5	5	4	<2	3	<2	270
Field_06	<2	9	8	52	98	408	15	<2	9	7	14	<2	3	<2	381
Slope_01	<2	10	13	19	669	86	11	9	4	4	<2	<2	<2	<2	277
Slope_02	8	10	10	13	358	67	9	12	3	7	<2	<2	<2	<2	248
Slope_03	11	18	11	17	505	69	9	16	4	10	2	<2	3	<2	679

(continued)

(continued)

Samples	(mg kg^{-1})														
	Se	As	Br	Rb	Sr	Zr	Pb	U	Th	Mo	Cd	Hg	Sn	Sb	Ba
Slope_04	<2	5	8	15	403	72	7	8	<2	<2	2	<2	<2	<2	1483
Slope_05	<2	16	10	25	570	115	11	14	2	4	<2	<2	3	<2	739
Slope_06	2	7	18	26	304	122	14	5	5	15	7	<2	2	<2	138
UR_01	<2	<2	6	3	1185	5	3	3	<2	<2	<2	<2	<2	<2	917
UR_02	<2	<2	13	<2	250	5	2	<2	<2	<2	<2	<2	<2	<2	147
UR_03	<2	2	24	<2	445	5	2	<2	<2	<2	<2	<2	<2	<2	195
UR_04	<2	<2	17	<2	226	6	1	<2	<2	<2	<2	<2	<2	<2	2427
UR_05	<2	<2	22	<2	351	4	3	3	<2	<2	<2	<2	<2	<2	135
UR_06	<2	<2	15	<2	435	4	3	3	<2	<2	<2	<2	<2	<2	108
MCM_01	<2	<2	14	4	585	8	5	8	2	<2	<2	<2	<2	<2	2326
MCM_02	<2	12	29	5	881	14	7	11	2	<2	<2	<2	<2	<2	51
MCM_03	4	9	11	8	1148	23	5	30	<2	13	<2	<2	<2	<2	73
MCM_04	<2	3	4	6	454	18	7	3	<2	<2	<2	<2	<2	<2	1848
MCM_05	20	12	3	7	1137	14	5	13	<2	8	<2	<2	<2	<2	105
MCM_06	<2	6	4	5	526	12	6	10	2	3	<2	<2	2	<2	31
ASL_01	25	10	2	11	684	43	7	11	<2	37	53	<2	<2	<2	56
ASL_02	<2	5	5	<2	217	8	2	13	<2	6	3	<2	<2	<2	89
ASL_03	<2	<2	13	3	314	6	6	8	2	<2	26	<2	3	<2	198
ASL_04	<2	16	10	2	720	8	6	11	<2	4	26	<2	<2	<2	27
ASL_05	<2	2	1	<2	273	8	2	7	<2	5	51	<2	<2	<2	22
ASL_06	<2	5	10	4	458	19	7	17	<2	7	63	<2	<2	6	3989
WD_01	<2	3	11	8	407	65	5	6	<2	<2	3	<2	<2	<2	1150
WD_02	<2	4	8	10	480	85	13	14	2	5	7	<2	4	<2	1528
WD_03	<2	7	9	16	580	134	12	9	2	2	5	<2	3	<2	942
								< below detection limit							

WD = Wadi Deposits

Soil, geology and sediment samples from the Wadi Al-Arab catchment

Samples	(%)								
	C	H	N	TOC	TIC	TC	Clay	Silt	Sand
Olive_01	8.75	0.79	0.02	1.52	7.13	8.64	40.26	43.85	15.90
Olive_02	9.93	0.72	0.01	1.88	7.92	9.79	20.85	39.16	39.99
Olive_03	8.38	0.75	0.00	0.89	7.12	8.00	48.03	42.63	9.34
Olive_04	9.32	0.74	0.01	1.57	7.66	9.22	26.26	55.23	18.51
Olive_05	10.14	0.44	0.00	0.84	8.98	9.82	26.96	45.57	27.47
Olive_06	5.05	1.39	0.09	2.39	2.38	4.76	35.53	51.15	13.32
Field_01	6.26	1.06	0.07	2.01	3.99	6.00	43.27	42.75	13.98
Field_02	8.38	0.84	0.04	1.76	6.55	8.31	32.36	49.06	18.57
Field_03	6.87	1.07	0.11	2.60	4.14	6.74	35.28	45.95	18.77

(continued)

(continued)

Samples	(%)								
	C	H	N	TOC	TIC	TC	Clay	Silt	Sand
Field_04	5.43	1.15	0.00	1.45	3.71	5.16	28.79	55.70	15.51
Field_05	8.66	0.90	0.05	2.06	6.41	8.47	27.52	44.82	27.67
Field_06	2.06	1.45	0.01	1.32	0.45	1.77	55.16	38.74	6.10
Slope_01	11.66	0.91	0.18	3.79	7.67	11.46	19.99	37.25	42.76
Slope_02	11.16	0.79	0.10	2.65	7.94	10.59	18.80	41.09	40.12
Slope_03	11.01	0.97	0.10	3.64	6.95	10.59	24.41	41.98	33.61
Slope_04	10.78	0.67	0.00	2.03	8.22	10.25	21.84	47.47	30.70
Slope_05	9.44	0.90	0.03	2.64	6.31	8.94	33.10	43.87	23.03
Slope_06	10.80	1.00	0.18	3.76	6.84	10.60	22.42	45.69	31.90
UR_01	11.86	0.18	0.00	0.50	10.90	11.40	No grain size analysis of the geology samples, because the rock samples were crushed		
UR_02	12.50	0.14	0.00	0.37	11.77	12.14			
UR_03	12.45	0.20	0.00	0.39	11.79	12.18			
UR_04	12.23	0.22	0.00	0.53	11.55	12.08			
UR_05	12.40	0.18	0.00	0.37	12.01	12.38			
UR_06	12.45	0.15	0.00	0.33	11.99	12.32			
MCM_01	12.07	0.17	0.00	0.30	11.43	11.73			
MCM_02	10.66	0.55	0.00	0.78	10.00	10.78			
MCM_03	10.84	0.26	0.00	0.51	9.98	10.49			
MCM_04	10.86	0.21	0.00	0.21	10.30	10.51			
MCM_05	11.31	0.18	0.00	0.30	10.57	10.87			
MCM_06	11.47	0.23	0.00	0.32	10.96	11.28			
ASL_01	21.48	2.12	0.11	15.42	6.11	21.52			
ASL_02	6.49	0.09	0.00	0.21	6.19	6.40			
ASL_03	11.67	0.33	0.00	0.31	11.16	11.46			
ASL_04	11.68	0.14	0.00	0.38	10.75	11.13			
ASL_05	7.92	0.09	0.00	0.15	7.60	7.75			
ASL_06	9.62	0.15	0.00	0.34	9.37	9.71			
WD_01	10.90	0.25	0.00	0.45	10.02	10.47	9.99	16.49	73.52
WD_02	8.95	0.62	0.02	1.20	7.70	8.90	19.49	57.06	23.45
WD_03	9.61	0.83	0.07	2.00	7.50	9.50	9.77	16.63	73.60
							<2 µm	<63 µm	<2 mm

WD = Wadi Deposits

Soil, geology and sediment samples from the Wadi Al-Arab catchment

Samples	(Bq kg^{-1})					
	U-238	E_U-238	Ra-226	E_Ra-226	Pb-210	E_Pb-210
Olive_01	25.73	1.49	29.07	1.33	43.47	1.30
Olive_02	37.44	2.13	159.48	5.04	125.86	3.27
Olive_03	45.75	1.33	37.33	1.51	42.11	1.43

(continued)

(continued)

Samples	(Bq kg^{-1})					
	U-238	E_U-238	Ra-226	E_Ra-226	Pb-210	E_Pb-210
Olive_04	13.89	0.78	17.12	1.05	19.15	1.05
Olive_05	22.24	0.51	22.11	0.70	38.93	0.90
Olive_06	40.40	1.90	49.45	1.54	69.01	2.76
Field_01	26.61	2.46	25.28	1.13	46.28	3.33
Field_02	69.08	2.63	87.96	3.50	91.73	6.97
Field_03	43.83	3.65	51.24	1.78	46.56	3.07
Field_04	57.14	2.79	54.64	1.94	56.29	3.10
Field_05	81.94	3.55	95.64	3.54	101.98	5.91
Field_06	30.61	2.20	49.72	2.24	61.19	4.16
Slope_01	95.06	10.17	107.69	3.94	111.62	5.25
Slope_02	113.90	4.44	101.48	3.22	103.32	8.16
Slope_03	170.80	8.51	192.50	5.18	169.89	9.85
Slope_04	79.83	2.39	75.97	2.90	78.95	3.87
Slope_05	166.03	1.68	129.56	2.71	131.59	5.92
Slope_06	33.01	3.60	41.36	1.74	74.90	3.07
UR_01	27.33	1.45	40.73	1.29	32.19	1.22
UR_02	20.15	1.09	18.48	0.97	14.51	1.03
UR_03	22.23	1.47	16.35	0.65	0.00	0.00
UR_04	5.14	0.38	8.29	0.47	9.47	0.81
UR_05	21.63	0.82	16.57	0.75	20.87	1.59
UR_06	21.19	0.70	17.08	0.82	18.87	1.17
MCM_01	56.25	2.48	58.36	2.68	45.18	2.35
MCM_02	106.15	4.78	71.82	2.19	57.46	2.93
MCM_03	365.54	14.99	317.40	8.75	245.51	9.82
MCM_04	18.05	0.96	24.79	1.39	-30.48	0.00
MCM_05	146.06	9.64	109.57	4.73	77.96	6.00
MCM_06	70.95	4.12	54.29	2.08	34.90	1.99
ASL_01	106.68	2.81	110.94	1.95	111.13	6.45
ASL_02	124.23	2.52	147.62	2.91	101.63	1.32
ASL_03	72.28	1.60	76.21	2.52	52.33	3.87
ASL_04	105.13	4.66	92.28	3.44	81.43	7.74
ASL_05	74.14	1.59	59.33	1.84	43.96	1.29
ASL_06	159.54	2.23	185.14	4.26	128.69	5.92
WD_01	54.61	1.69	47.30	1.53	47.28	3.40
WD_02	134.91	3.24	146.18	4.61	115.14	3.22
WD_03	111.98	2.80	116.38	2.72	115.42	2.48

WD = Wadi Deposits
E_ = error

Soil, geology and sediment samples from the Wadi Al-Arab catchment

Samples	(Bq kg^{-1})					
	Th-232	E_Th-232	Th-228	E_Th-228	K-40	E_K-40
Olive_01	13.38	0.62	12.39	0.57	97.85	5.68
Olive_02	13.52	1.41	12.00	0.94	126.45	5.31
Olive_03	15.70	1.22	15.53	1.00	129.84	6.10
Olive_04	7.83	0.67	7.53	0.60	67.14	3.36
Olive_05	15.09	0.85	16.05	0.66	125.45	3.64
Olive_06	38.49	2.34	37.43	2.18	194.68	7.01
Field_01	26.55	1.78	24.71	1.01	173.63	5.04
Field_02	17.10	1.12	17.17	0.91	90.86	3.18
Field_03	28.14	1.66	28.01	1.14	203.71	9.17
Field_04	25.68	1.10	26.94	0.93	161.65	3.56
Field_05	16.45	1.38	16.73	0.85	140.56	6.18
Field_06	40.05	1.99	39.56	1.23	168.69	4.89
Slope_01	10.87	0.74	11.09	0.69	102.43	4.51
Slope_02	11.82	0.83	11.22	0.68	59.78	2.81
Slope_03	11.77	1.11	13.02	0.81	66.42	3.59
Slope_04	7.64	0.61	9.93	0.78	64.20	5.59
Slope_05	17.33	1.05	17.24	0.70	93.51	3.68
Slope_06	14.61	0.77	14.26	0.83	82.24	4.19
UR_01	1.14	0.18	1.80	0.14	10.40	0.46
UR_02	1.63	0.28	0.65	0.07	4.11	0.34
UR_03	1.05	0.29	1.18	0.15	0.17	0.02
UR_04	1.23	0.22	1.49	0.14	7.24	0.43
UR_05	0.97	0.16	0.71	0.09	1.40	0.12
UR_06	0.74	0.11	0.82	0.07	4.07	0.26
MCM_01	0.28	0.06	1.57	0.06	2.89	0.29
MCM_02	4.40	0.70	3.65	0.19	21.91	1.56
MCM_03	4.82	0.76	5.66	0.39	32.48	2.47
MCM_04	3.67	0.76	5.11	0.51	27.05	2.03
MCM_05	6.61	1.34	3.32	0.63	39.39	3.47
MCM_06	2.65	0.58	2.53	0.21	17.17	1.53
ASL_01	7.09	0.75	6.34	0.39	151.13	4.55
ASL_02	2.95	0.29	1.27	0.06	8.99	0.46
ASL_03	1.14	0.23	1.14	0.08	4.52	0.38
ASL_04	1.74	0.37	1.98	0.19	4.76	0.40
ASL_05	1.87	0.25	1.60	0.14	7.72	0.66
ASL_06	1.66	0.23	2.42	0.14	0.00	0.00
WD_01	5.55	0.59	5.38	0.30	45.98	2.90
WD_02	7.46	0.80	9.99	0.64	48.81	4.20
WD_03	14.59	1.27	14.45	1.00	109.93	4.18

WD = Wadi Deposits
E_ = error

Lake sediment samples from Wadi Al-Arab reservoir

Samples	(Bq kg^{-1})	
	Cs-137	E_Cs-137
Olive_01	21.06	0.69
Olive_02	14.79	1.40
Olive_03	13.96	1.07
Olive_04	3.90	0.50
Olive_05	24.78	1.09
Olive_06	34.21	2.19
Field_01	12.53	0.53
Field_02	9.14	0.60
Field_03	9.63	0.54
Field_04	8.50	0.55
Field_05	10.66	0.45
Field_06	8.50	0.59
Slope_01	30.72	1.03
Slope_02	29.88	2.09
Slope_03	38.33	2.00
Slope_04	16.12	1.14
Slope_05	23.83	1.46
Slope_06	38.18	1.18
UR_01	0.00	0.00
UR_02	0.00	0.00
UR_03	0.00	0.00
UR_04	0.00	0.00
UR_05	0.00	0.00
UR_06	0.00	0.00
MCM_01	0.00	0.00
MCM_02	0.00	0.00
MCM_03	0.00	0.00
MCM_04	0.00	0.00
MCM_05	0.00	0.00
MCM_06	0.00	0.00
ASL_01	0.00	0.00
ASL_02	0.00	0.00
ASL_03	0.00	0.00
ASL_04	0.00	0.00
ASL_05	0.00	0.00
ASL_06	0.00	0.00
WD_01	0.00	0.00
WD_02	5.72	0.85
WD_03	7.84	0.65

WD = Wadi Deposits
E_ = error

Lake sediment samples from Wadi Al-Arab reservoir

Lake Corer_depth	(mg kg^{-1})										
	LOI	Na	Mg	Al	Si	P	S	K	Ca	Ti	V
L1_5 cm	321,800	1020	7320	32,290	90,960	5870	1943	3960	240,820	2420	118
L1_10 cm	324,600	1060	6750	31,430	86,750	7300	2367	3640	246,910	2210	120
L1_15 cm	315,700	1020	7500	33,360	91,690	6790	689	3910	242,520	2280	134
L1_20 cm	325,800	1010	7130	30,910	88,480	7000	622	3760	247,780	2160	108
L1_25 cm	326,200	990	7430	32,540	92,660	6710	1619	3960	238,130	2240	115
L1_30 cm	322,800	930	6770	32,870	93,850	7050	693	3920	241,690	2270	120
L3_5 cm	333,700	1030	7720	32,860	91,410	4750	2622	3930	236,870	2520	94
L3_10 cm	324,100	930	6800	31,920	91,450	5570	831	3890	242,900	2430	109
L3_15 cm	339,400	980	6490	28,650	85,820	6140	725	3630	250,830	2220	119
L3_20 cm	336,200	1030	6980	27,310	86,030	6470	2274	3410	248,950	2160	120
L3_25 cm	335,400	880	7340	31,060	87,280	5980	689	3700	249,300	2200	110
L3_30 cm	328,100	910	6910	31,960	88,340	6490	615	3760	247,360	2160	115
L3_35 cm	325,100	1150	7020	26,770	88,150	6890	981	3500	253,650	2110	103
L3_42 cm	324,200	990	7010	31,520	90,790	6610	1248	3790	243,500	2250	127
L4_5 cm	329,400	1000	7670	33,240	92,450	4950	2741	3980	237,100	2510	113
L4_10 cm	336,000	960	6580	28,890	87,170	6020	777	3640	251,640	2300	112
L4_15 cm	335,000	1090	6710	26,930	84,700	6440	1290	3520	252,990	2210	101
L4_20 cm	331,200	960	7260	31,450	88,660	6520	1559	3680	245,400	2280	112
L4_25 cm	315,400	960	7340	33,210	91,280	6550	648	3910	243,290	2270	115
L4_30 cm	333,200	1100	6870	27,780	86,680	7040	789	3530	249,230	2080	105
L4_35 cm	334,700	990	7590	29,810	87,040	6180	1835	3770	243,240	2110	106
L4_39 cm	321,500	950	6860	33,050	93,820	6880	922	3910	238,500	2310	128
L5_5 cm	334,900	1030	7290	30,450	87,620	5490	2356	3690	244,200	2330	115
L5_10 cm	330,200	1180	6550	25,950	93,700	6310	919	3470	248,860	2430	99
L5_15 cm	350,400	960	6870	25,600	79,140	5990	2681	3160	255,730	2000	109
L5_20 cm	328,700	1040	6760	28,700	85,530	6910	682	3490	253,710	2130	128
L5_25 cm	330,100	1020	6930	28,940	84,810	6740	883	3580	250,460	2090	122
L5_30 cm	334,000	990	6960	28,000	84,270	6470	1516	3500	251,610	2050	109
L5_35 cm	317,500	1030	6170	29,690	90,940	7840	688	3490	248,800	2340	121
L5_40 cm	324,500	990	6640	30,860	88,940	7390	979	3630	246,200	2250	118

Lake sediment samples from Wadi Al-Arab reservoir

Lake Corer_depth	(mg kg^{-1})												
	Cr	Mn	Fe	Co	Ni	Cu	Zn	Se	As	Br	Rb	Sr	Zr
L1_5 cm	211	335	20,980	<10	91	52	227	3	8	12	26	611	80
L1_10 cm	247	305	20,630	<10	109	61	243	5	10	11	24	730	73
L1_15 cm	252	331	20,840	<10	107	56	224	3	9	9	25	721	73
L1_20 cm	250	319	19,440	<10	105	54	228	3	8	9	24	722	77
L1_25 cm	240	320	21,120	<10	108	55	231	4	9	11	25	737	72
L1_30 cm	251	332	20,420	<10	108	58	244	2	8	9	26	695	76
L3_5 cm	175	316	22,240	<10	78	46	212	3	7	16	25	555	80

(continued)

(continued)

Lake Corer_depth	(mg kg^{-1})												
	Cr	Mn	Fe	Co	Ni	Cu	Zn	Se	As	Br	Rb	Sr	Zr
L3_10 cm	197	328	21,120	<10	78	48	223	<2	7	8	23	543	80
L3_15 cm	205	302	19,230	<10	86	52	230	2	7	8	21	593	82
L3_20 cm	203	314	19,580	<10	96	54	220	4	7	11	19	707	84
L3_25 cm	209	328	19,720	<10	98	51	223	2	8	8	23	649	71
L3_30 cm	241	303	20,000	<10	99	59	232	2	9	7	24	678	69
L3_35 cm	215	336	18,540	<10	98	51	215	3	8	9	19	718	81
L3_42 cm	228	301	20,560	<10	103	57	241	4	9	10	24	708	75
L4_5 cm	189	326	23,080	<10	82	47	216	2	8	14	26	574	81
L4_10 cm	204	315	19,370	<10	84	49	223	<2	7	8	21	580	85
L4_15 cm	211	316	18,610	<10	89	53	227	3	7	9	19	638	88
L4_20 cm	229	302	20,840	<10	108	58	231	2	10	10	24	699	73
L4_25 cm	241	348	20,980	<10	104	58	222	2	8	9	25	707	72
L4_30 cm	233	327	18,260	<10	98	55	226	3	7	10	20	716	79
L4_35 cm	228	313	20,210	<10	101	56	233	3	8	11	23	744	69
L4_39 cm	241	304	20,700	<10	107	57	246	2	9	9	26	695	75
L5_5 cm	198	310	20,770	<10	87	52	216	3	8	13	24	622	84
L5_10 cm	184	326	18,330	<10	79	47	204	<2	7	7	19	599	129
L5_15 cm	196	288	17,910	<10	87	50	209	4	7	10	17	719	76
L5_20 cm	234	311	18,470	<10	103	57	228	2	9	8	22	697	82
L5_25 cm	242	313	19,090	<10	103	59	245	2	9	8	20	724	71
L5_30 cm	224	286	19,020	<10	100	55	233	4	8	9	23	746	70
L5_35 cm	256	294	20,070	<10	103	63	252	2	10	9	23	706	95
L5_40 cm	256	321	19,930	<10	108	61	257	3	9	9	24	720	85
				< below detection limit									

Lake sediment samples from Wadi Al-Arab reservoir

Lake Corer_depth	(mg kg^{-1})												
	Cr	Mn	Fe	Co	Ni	Cu	Zn	Se	As	Br	Rb	Sr	Zr
L1_5 cm	211	335	20,980	<10	91	52	227	3	8	12	26	611	80
L1_10 cm	247	305	20,630	<10	109	61	243	5	10	11	24	730	73
L1_15 cm	252	331	20,840	<10	107	56	224	3	9	9	25	721	73
L1_20 cm	250	319	19,440	<10	105	54	228	3	8	9	24	722	77
L1_25 cm	240	320	21,120	<10	108	55	231	4	9	11	25	737	72
L1_30 cm	251	332	20,420	<10	108	58	244	2	8	9	26	695	76
L3_5 cm	175	316	22,240	<10	78	46	212	3	7	16	25	555	80
L3_10 cm	197	328	21,120	<10	78	48	223	<2	7	8	23	543	80
L3_15 cm	205	302	19,230	<10	86	52	230	2	7	8	21	593	82
L3_20 cm	203	314	19,580	<10	96	54	220	4	7	11	19	707	84
L3_25 cm	209	328	19,720	<10	98	51	223	2	8	8	23	649	71
L3_30 cm	241	303	20,000	<10	99	59	232	2	9	7	24	678	69

(continued)

(continued)

Lake Corer_depth	(mg kg^{-1})												
	Cr	Mn	Fe	Co	Ni	Cu	Zn	Se	As	Br	Rb	Sr	Zr
L3_35 cm	215	336	18,540	<10	98	51	215	3	8	9	19	718	81
L3_42 cm	228	301	20,560	<10	103	57	241	4	9	10	24	708	75
L4_5 cm	189	326	23,080	<10	82	47	216	2	8	14	26	574	81
L4_10 cm	204	315	19,370	<10	84	49	223	<2	7	8	21	580	85
L4_15 cm	211	316	18,610	<10	89	53	227	3	7	9	19	638	88
L4_20 cm	229	302	20,840	<10	108	58	231	2	10	10	24	699	73
L4_25 cm	241	348	20,980	<10	104	58	222	2	8	9	25	707	72
L4_30 cm	233	327	18,260	<10	98	55	226	3	7	10	20	716	79
L4_35 cm	228	313	20,210	<10	101	56	233	3	8	11	23	744	69
L4_39 cm	241	304	20,700	<10	107	57	246	2	9	9	26	695	75
L5_5 cm	198	310	20,770	<10	87	52	216	3	8	13	24	622	84
L5_10 cm	184	326	18,330	<10	79	47	204	<2	7	7	19	599	129
L5_15 cm	196	288	17,910	<10	87	50	209	4	7	10	17	719	76
L5_20 cm	234	311	18,470	<10	103	57	228	2	9	8	22	697	82
L5_25 cm	242	313	19,090	<10	103	59	245	2	9	8	20	724	71
L5_30 cm	224	286	19,020	<10	100	55	233	4	8	9	23	746	70
L5_35 cm	256	294	20,070	<10	103	63	252	2	10	9	23	706	95
L5_40 cm	256	321	19,930	<10	108	61	257	3	9	9	24	720	85
				< below Detection Limit									

Lake sediment samples from Wadi Al-Arab reservoir

Lake Corer_depth	(%)			(Bq kg^{-1})			
	Clay	Silt	Sand	U-238	E_U-238	Ra-226	E_Ra-226
L1_5 cm				124.42	2.53	165.64	6.20
L1_10 cm				177.40	4.52	196.07	6.46
L1_15 cm				165.20	4.11	162.78	5.28
L1_20 cm				162.20	3.84	155.33	5.60
L1_25 cm				148.48	3.56	158.35	5.11
L1_30 cm				162.26	9.30	151.08	5.56
L3_5 cm				110.41	4.31	148.01	4.60
L3_10 cm				131.73	7.64	122.59	4.96
L3_15 cm				129.43	4.79	134.21	3.92
L3_20 cm				152.61	2.75	165.10	5.89
L3_25 cm				151.79	7.44	169.53	6.32
L3_30 cm				178.20	6.42	187.76	7.24
L3_35 cm				145.62	5.24	157.31	5.02
L3_42 cm				168.94	3.40	167.55	4.40
L4_5 cm	46.86	52.66	0.48	103.53	7.04	150.34	6.44
L4_10 cm	39.9	60.01	0.09	121.64	2.68	131.62	2.87
L4_15 cm	43.34	56.39	0.26	122.43	4.53	144.27	3.49

(continued)

(continued)

Lake Corer_depth	(%)			(Bq kg^{-1})			
	Clay	Silt	Sand	U-238	E_U-238	Ra-226	E_Ra-226
L4_20 cm	50.25	49.67	0.08	163.03	3.59	160.73	3.41
L4_25 cm	47.04	52.79	0.18	153.96	7.08	150.36	4.99
L4_30 cm	47.23	51.81	0.96	154.36	5.09	159.67	3.85
L4_35 cm	48.92	51.00	0.08	137.48	5.77	156.29	5.30
L4_39 cm	45.03	54.73	0.24	184.11	6.08	164.47	7.06
L5_5 cm				124.94	3.00	164.01	5.75
L5_10 cm				133.83	5.62	195.49	4.70
L5_15 cm				163.53	8.34	195.24	4.39
L5_20 cm				145.18	4.21	196.29	4.42
L5_25 cm				171.35	8.57	203.38	5.40
L5_30 cm				164.60	10.53	203.76	8.05
L5_35 cm				197.95	9.50	226.75	6.24
L5_40 cm				162.78	3.09	193.50	3.23
	<2 μm	<63 μm	<2 mm				

Lake sediment samples from Wadi Al-Arab reservoir

Lake Corer_depth	(Bq kg^{-1})					
	Pb-210	E_Pb-210	Th-232	E_Th-232	Th-228	E_Th-228
L1_5 cm	135.67	7.14	17.01	1.78	16.64	0.86
L1_10 cm	174.11	4.16	0.00	0.00	15.99	1.06
L1_15 cm	152.62	3.72	15.24	1.33	16.01	0.82
L1_20 cm	144.57	4.28	19.29	1.82	16.33	1.25
L1_25 cm	146.42	3.62	16.22	1.53	14.92	1.35
L1_30 cm	158.80	8.89	17.69	1.49	15.05	1.00
L3_5 cm	154.87	5.73	17.85	1.61	14.23	0.69
L3_10 cm	129.50	5.44	0.00	0.00	15.36	0.87
L3_15 cm	132.77	2.26	14.51	1.19	14.13	0.50
L3_20 cm	148.49	2.82	14.26	1.70	14.80	0.72
L3_25 cm	162.87	6.51	13.10	1.09	14.08	1.11
L3_30 cm	185.67	5.76	16.30	1.91	16.55	1.10
L3_35 cm	155.67	10.45	12.47	0.81	14.65	0.87
L3_42 cm	165.77	6.00	15.52	3.91	16.62	1.01
L4_5 cm	137.55	3.71	15.91	1.65	16.03	1.18
L4_10 cm	139.82	5.73	14.71	1.40	14.44	0.70
L4_15 cm	124.03	5.71	12.98	2.83	12.82	0.68
L4_20 cm	165.72	7.46	14.80	1.12	15.70	0.74
L4_25 cm	150.08	8.55	16.03	1.17	14.19	0.81
L4_30 cm	143.20	13.32	18.63	1.34	14.08	0.81

(continued)

(continued)

Lake Corer_depth	(Bq kg^{-1})					
	Pb-210	E_Pb-210	Th-232	E_Th-232	Th-228	E_Th-228
L4_35 cm	140.02	6.30	15.16	3.50	10.29	0.77
L4_39 cm	164.83	13.35	15.14	1.80	15.12	1.24
L5_5 cm	142.75	4.28	19.75	1.30	18.51	0.98
L5_10 cm	134.66	5.93	16.05	0.88	15.18	1.08
L5_15 cm	169.82	5.04	12.82	1.62	14.79	0.65
L5_20 cm	166.06	5.48	13.34	0.94	14.83	0.54
L5_25 cm	178.36	6.24	14.92	1.84	14.81	1.09
L5_30 cm	170.63	12.12	0.00	0.00	15.68	1.26
L5_35 cm	211.24	13.10	16.53	1.31	16.12	1.24
L5_40 cm	169.45	8.98	14.74	1.10	14.45	0.63

Lake sediment samples from Wadi Al-Arab reservoir

Lake Corer_depth	(Bq kg^{-1})			
	K-40	E_K-40	Cs-137	E_Cs-137
L1_5 cm	120.90	4.28	2.79	0.43
L1_10 cm	103.85	5.25	3.49	0.41
L1_15 cm	124.41	5.57	3.50	0.44
L1_20 cm	121.68	7.26	3.56	0.59
L1_25 cm	123.03	4.79	0.00	0.00
L1_30 cm	119.77	6.59	4.88	0.49
L3_5 cm	133.68	3.74	0.95	0.10
L3_10 cm	113.10	6.90	0.00	0.00
L3_15 cm	110.13	3.96	2.98	0.25
L3_20 cm	107.90	5.18	1.99	0.31
L3_25 cm	118.14	4.61	5.21	0.72
L3_30 cm	128.34	5.52	5.20	0.50
L3_35 cm	102.01	5.04	11.70	1.43
L3_42 cm	120.95	4.12	13.94	1.46
L4_5 cm	105.60	9.19	0.00	0.00
L4_10 cm	117.33	3.75	8.21	1.22
L4_15 cm	101.42	5.38	8.19	0.54
L4_20 cm	120.42	3.25	11.16	1.00
L4_25 cm	129.27	7.50	12.22	1.28
L4_30 cm	103.46	3.62	10.65	1.12
L4_35 cm	118.40	6.87	13.80	2.08
L4_39 cm	140.98	12.12	0.00	0.00
L5_5 cm	99.59	4.08	2.55	0.39

(continued)

(continued)

Lake Corer_depth	(Bq kg^{-1})			
	K-40	E_K-40	Cs-137	E_Cs-137
L5_10 cm	110.62	4.98	3.45	0.28
L5_15 cm	92.33	4.52	3.73	0.39
L5_20 cm	111.72	3.24	4.40	0.20
L5_25 cm	106.65	4.91	6.23	0.62
L5_30 cm	115.35	8.19	5.94	0.93
L5_35 cm	101.42	4.56	5.23	0.58
L5_40 cm	105.84	3.49	15.87	0.95

Water samples from the Wadi Al-Arab reservoir

Sample_depth	X	Y	pH	Temp (°C)	EC (μS cm^{-1})	Redox (mV)
A_4	74 774.18	3 612 113.27	8.08	25.90	1033.00	51
A_8			8.31	25.60	1031.00	132
A_10			7.53	26.00	1042.00	8
A_12			7.36	25.70	1073.00	−49
A_14			7.28	24.30	1078.00	−36
A_16			7.33	24.50	1077.00	−45
B_2	747 667.51	3 612 082.51	8.19	26.50	1034.00	130
B_4			8.20	25.80	1036.00	102
B_8			8.01	26.40	1034.00	47
B_10			7.45	24.50	1055.00	−25
B_12			7.27	24.10	1072.00	−42
B_14			7.31	23.10	1083.00	−39
B_17			7.26	23.30	1079.00	−36

Sample_depth	(mg l^{-1})					
	Ammonium	PO_4	Nitrate	Nitrite	Gesamt P	Bromide
A_4	<0.01	<0.07	0.09	<0.01	<0.1	0.90
A_8	0.03	<0.07	0.27	<0.01	<0.1	0.94
A_10	0.72	0.13	0.35	<0.01	<0.1	0.89
A_12	6.3	1.24	0.73	<0.01	0.4661	0.76
A_14	7	1.4	0.46	<0.01	0.554	0.77
A_16	7.2	1.44	0.42	<0.01	0.509	0.77
B_2	<0.01	<0.07	0.27	<0.01	<0.1	0.89
B_4	<0.01	<0.07	0.13	<0.01	<0.1	0.90
B_8	0.07	<0.07	0.29	<0.01	<0.1	0.92
B_10	3.5	0.86	0.2	<0.01	0.2669	0.81

(continued)

(continued)

Sample_depth	(mg l^{-1})					
	Ammonium	PO_4	Nitrate	Nitrite	Gesamt P	Bromide
B_12	6.4	1.45	0.07	<0.01	0.493	0.73
B_14	7	1.24	0.35	<0.01	0.513	0.74
B_17	7.5	1.45	0.35	<0.01	0.564	0.79

Water samples from the Wadi Al-Arab reservoir

Sample_depth	(ng mL^{-1})						
	As	Cd	Cs	Fe	Mn	Rb	U
A_4	2.3	0.075	0.03	32	14.2	5.3	3
A_8	2.37	0.09	0.0243	34	3	5.3	3.2
A_10	2.34	0.067	0.036	121	156	5.2	2.8
A_12	1.4	0.017	0.066	101	236	5.1	1.92
A_14	1.5	0.0172	0.073	131	270	5.2	1.84
A_16	1.42	0.0175	0.075	131	280	5.2	1.81
B_2	2.29	0.079	0.031	22.8	1.27	5.3	3.1
B_4	2.32	0.072	0.031	33	2.38	5.2	3
B_8	2.27	0.082	0.032	32	22.6	5.2	3.2
B_10	2.3	0.031	0.052	135	450	5.2	2.43
B_12	1.55	0.0204	0.063	75	249	5	2
B_14	1.62	0.029	0.064	87	268	5.1	1.93
B_17	1.38	0.0129	0.072	90	270	5.2	1.82

Sample_depth	(mg l^{-1})					
	B	Ba	Ca	K	Li	Mg
A_4	0.1	0.2	53.9	8.69	<0.07	39.86
A_8	0.1	0.2	54.3	8.68	<0.07	40.26
A_10	0.1	0.2	59	8.6	<0.07	39.65
A_12	0.1	0.3	83.6	8.5	<0.07	36.59
A_14	0.1	0.3	85.2	8.56	<0.07	37.23
A_16	0.1	0.3	85	8.5	<0.07	37.04
B_2	0.1	0.1	55.1	8.64	<0.07	39.55
B_4	0.1	0.1	55.2	8.6	<0.07	39.67
B_8	0.1	0.1	55.9	8.62	<0.07	39.82
B_10	0.1	0.3	74.8	8.46	<0.07	37.34
B_12	0.1	0.4	84.7	8.53	<0.07	36.74
B_14	0.1	0.3	85.4	8.65	<0.07	37.08
B_17	0.1	0.4	86.2	8.62	<0.07	37.28

Water samples from the Wadi Al-Arab reservoir

Sample_depth	($mg\ l^{-1}$)					
	Na	Si	Sr	Chloride	Nitrate	Sulfate
A_4	113.3	8.48	0.907	197	<0.04	111.00
A_8	114.3	8.55	0.915	201	<0.04	114.00
A_10	110	8.57	0.936	194	<0.04	108.00
A_12	97.8	8.63	1.027	167	<0.04	75.40
A_14	97.2	8.69	1.046	167	<0.04	74.60
A_16	96.8	8.68	1.047	168	<0.04	74.40
B_2	112.2	8.5	0.911	192	<0.04	107.00
B_4	112.3	8.5	0.93	192	<0.04	107.00
B_8	111.9	8.5	0.921	197	<0.04	111.00
B_10	101.7	8.58	1.001	178	<0.04	86.80
B_12	97.3	8.62	1.037	161	<0.04	75.00
B_14	98.2	8.69	1.039	162	<0.04	72.90
B_17	97.1	8.8	1.054	168	<0.04	74.50

($ng\ l^{-1}$)	Sample_depth	
	B_2	B_14
Trifluralin	nn	0.13
t-NP	nn	88.4
Coffein	6.4	53.2
Galaxolide	2.9	9.17
Tonalide	0.8	1.54
Triclosan	0.3	1.15
BPA	24.6	61.2
Carbamazepin	nn	21.1
Hexazinone	nn	5.14

Appendix Chapter 6

S. Kraushaar, *Soil Erosion and Sediment Flux in Northern Jordan*,
Springer Theses, DOI 10.1007/978-3-319-31888-2

Soil map parameters

Profile	Horizon depth [cm]	Geology	Position	Profile depth [cm]	Northing	Easting	Elevation
P01	50.0	ASL	Foot Slope	185+	3612177.2	756539.4	162.1
P01	70.0	ASL	Foot Slope	185+	3612177.2	756539.4	162.1
P01	185.0	ASL	Foot Slope	185+	3612177.2	756539.4	162.1
P02	43.0	ASL	Slope	60	3612141.1	756472.3	205.1
P03	16.0	MCM	Slope	16	3612056.0	756268.8	304.5
P04	10.0	MCM	Upslope	78	3613519.4	761295.0	361.9
P04	71.0	MCM	Upslope	78	3613519.4	761295.0	361.9
P05	75.0	MCM	Saddle	100	3613350.5	762696.7	400.9
P06	120.0	MCM	Slope	133	3611887.9	749745.8	18.7
P07	30.0	MCM	Saddle	99+	3611979.6	749819.0	-0.8
P07	90.0	MCM	Saddle	99+	3611979.6	749819.0	-0.8
P08	12.0	MCM	Upslope	47	3611962.9	764654.8	422.7
P09	18.0	UR	Upslope	180	3614985.3	765759.5	580.1
P09	30.0	UR	Upslope	180	3614985.3	765759.5	580.1
P09	150.0	UR	Upslope	180	3614985.3	765759.5	580.1
P10	10.0	ASL	Slope	135	3607192.8	763738.3	469.0
P10	35.0	ASL	Slope	135	3607192.8	763738.3	469.0
P10	85.0	ASL	Slope	135	3607192.8	763738.3	469.0
P11	35.0	BASALT	Upslope	46	3615273.9	749795.2	316.2
P12	30.0	ASL	Upslope	93	3600432.6	762936.6	653.9
P13	7.0	ASL	Saddle	156	3605401.8	762999.9	519.0
P14	45.0	ASL	Alluvium	318+	3612721.8	755000.2	108.5
P14	53.0	ASL	Alluvium	318+	3612721.8	755000.2	108.5
P14	153.0	ASL	Alluvium	318+	3612721.8	755000.2	108.5
P15	30.0	MCM	Colluvium	225	3612931.9	753808.7	119.1
P15	125.0	MCM	Colluvium	225	3612931.9	753808.7	119.1
P15	195.0	MCM	Colluvium	225	3612931.9	753808.7	119.1
P16	13.0	MCM	Slope	13	3608498.1	757138.9	338.4
P17	14.0	ASL	Alluvium	145	3608206.4	752494.8	116.9
P17	145.0	ASL	Alluvium	145	3608206.4	752494.8	116.9
P18	27.0	UR	Slope	27	3614871.3	754611.8	330.2
P19	33.0	MCM	Colluvium	101+	3614256.7	758388.2	243.1
P20	19.0	ASL	Colluvium	150+	3603874.4	756949.6	348.9
P20	33.0	ASL	Colluvium	150+	3603874.4	756949.6	348.9
P21	30.0	MCM	Colluvium	80	3612749.0	749142.4	-82.0
P21	38.0	MCM	Colluvium	80	3612749.0	749142.4	-82.0
P22	60.0	MCM	Slope	60	3612688.0	749196.3	-63.0
P23	12.0	MCM	Slope	12	3612927.3	757325.0	317.0
P24	30.0	MCM	Slope	30	3612201.0	757158.3	315.7
P25	20.0	MCM	Uplsope	20	3611894.6	757678.2	369.8
P26	15.0	MCM	Saddle	30	3612029.9	757282.7	322.4
P26	30.0	MCM	Saddle	30	3612029.9	757282.7	322.4
P27	50.0	MCM	Uplsope	207	3608343.1	755725.1	337.1
P27	90.0	MCM	Uplsope	207	3608343.1	755725.1	337.1
P27	127.0	MCM	Uplsope	207	3608343.1	755725.1	337.1
P27	207.0	MCM	Uplsope	207	3608343.1	755725.1	337.1
P27	278.0	MCM	Uplsope	207	3608343.1	755725.1	337.1
P28	23.0	Caliche	Uplsope	110	3610065.3	765850.7	446.2
P28	55.0	Caliche	Upslope	110	3610065.3	765850.7	446.2

(shaded)	Profiles used to estimate variability of soils on different morphological positions
(white)	Leading profiles, fundaments of the soil map.

Profile	Bulk density [g cm-3]	Kfs [cm h-1]	Infiltration class (Landon, 1991)	K-Factor (Loch and Rosewell, 1992)	Field capacity [g]
P01	1.15	4.0	3.0	0.1	29.5
P01	n.d.	4.0	3.0	0.1	n.d.
P01	n.d.	4.0	3.0	0.1	n.d.
P02	n.d.	0.1	1.0	0.1	n.d.
P03	0.91	2.6	3.0	0.0	30.9
P04	1.11	0.7	1.0	0.1	38.5
P04	n.d.	0.7	1.0	0.1	n.d.
P05	1.34	1.9	2.0	0.1	14.9
P06	1.36	0.8	1.0	0.1	18.7
P07	1.20	1.4	2.0	0.1	28.0
P07	1.34	1.4	2.0	0.1	29.3
P08	1.08	2.3	3.0	0.1	33.9
P09	1.05	0.5	1.0	0.1	35.9
P09	n.d.	0.5	1.0	0.1	n.d.
P09	n.d.	0.5	1.0	0.1	n.d.
P10	n.d.	0.5	1.0	0.1	n.d.
P10	n.d.	0.5	1.0	0.1	n.d.
P10	n.d.	0.5	1.0	0.1	n.d.
P11	1.12	2.4	3.0	0.1	25.3
P12	1.01	2.6	3.0	0.1	25.2
P13	n.d.	0.5	1.0	0.1	n.d.
P14	1.01	1.2	2.0	0.1	21.9
P14	n.d.	1.2	2.0	0.1	n.d.
P14	n.d.	1.2	2.0	0.1	n.d.
P15	0.99	1.1	2.0	0.1	21.2
P15	n.d.	1.1	2.0	0.1	n.d.
P15	n.d.	1.1	2.0	0.1	n.d.
P16	0.94	2.6	3.0	0.0	27.0
P17	n.d.	0.9	2.0	0.1	n.d.
P17	n.d.	0.9	2.0	0.1	n.d.
P18	1.05	2.3	3.0	0.1	22.4
P19	1.25	0.1	1.0	0.1	19.5
P20	n.d.	1.5	2.0	0.1	n.d.
P20	n.d.	1.5	2.0	0.1	n.d.
P21	n.d.	0.8	1.0	0.1	n.d.
P21	n.d.	0.8	1.0	0.1	n.d.
P22	n.d.	0.8	1.0	0.1	n.d.
P23	n.d.	2.6	3.0	0.1	n.d.
P24	n.d.	2.6	3.0	0.0	n.d.
P25	n.d.	0.8	1.0	0.1	n.d.
P26	n.d.	1.9	2.0	0.1	n.d.
P26	n.d.	1.9	2.0	0.1	n.d.
P27	n.d.	2.3	3.0	0.1	n.d.
P27	n.d.	2.3	3.0	0.1	n.d.
P27	n.d.	2.3	3.0	0.1	n.d.
P27	n.d.	2.3	3.0	0.1	n.d.
P27	n.d.	2.3	3.0	0.1	n.d.
P28	n.d.	2.6	3.0	0.1	n.d.
P28	n.d.	2.6	3.0	0.1	n.d.

n.d. = No data, coring sleves were not possible to extract due to the high stone content.

Profile	[%]						
	Field capacity	Skelett	Clay	Silt	Sand	Fine mat.	Coarser mat.
P01	20.5	35.0	35.1	61.7	3.2	99.5	0.5
P01	n.d.	42.5	37.5	58.9	3.6	98.9	1.1
P01	n.d.	50.0	43.4	54.7	1.9	99.8	0.2
P02	n.d.	n.d.	31.5	64.1	4.4	99.1	0.9
P03	25.2	15.0	18.2	74.0	7.8	98.6	1.4
P04	25.8	25.0	30.9	60.3	8.8	98.4	1.6
P04	n.d.	3.0	39.1	57.4	3.5	99.7	0.3
P05	10.0	n.d.	28.1	62.4	9.5	96.1	3.9
P06	12.1	n.d.	31.5	41.4	27.1	82.6	17.4
P07	18.8	30.0	30.1	60.4	9.5	96.6	3.4
P07	17.9	3.0	43.0	50.0	7.0	97.9	2.1
P08	23.9	15.0	37.5	57.8	4.7	99.1	0.9
P09	25.4	4.0	46.2	49.8	3.9	99.5	0.5
P09	n.d.	4.0	27.1	70.1	2.8	99.8	0.2
P09	n.d.	2.0	44.7	52.8	2.5	99.8	0.2
P10	n.d.	4.0	29.6	59.7	10.7	96.7	3.3
P10	n.d.	80.0	39.6	50.9	9.5	96.6	3.4
P10	n.d.	70.0	40.5	54.0	5.5	99.0	1.0
P11	18.4	3.0	16.9	54.8	28.4	85.4	14.6
P12	20.0	45.0	34.4	60.3	5.3	99.3	0.7
P13	n.d.	45.0	25.2	62.1	12.7	95.8	4.2
P14	17.9	35.0	41.8	51.8	6.4	98.5	1.5
P14	n.d.	85.0	40.7	48.6	10.7	96.0	4.0
P14	n.d.	85.0	37.8	49.8	12.4	95.2	4.8
P15	17.6	n.d.	29.8	62.4	7.8	98.5	1.5
P15	n.d.	85.0	37.2	58.7	4.1	99.6	0.4
P15	n.d.	70.0	42.3	53.8	3.8	99.7	0.3
P16	22.3	20.0	24.4	68.1	7.5	98.4	1.6
P17	n.d.	20.0	20.9	58.1	21.1	90.9	9.1
P17	n.d.	85.0	25.9	53.5	20.6	89.4	10.6
P18	17.6	60.0	26.3	66.1	7.5	98.7	1.3
P19	13.4	25.0	30.7	59.5	9.8	98.0	2.1
P20	n.d.	40.0	17.9	63.4	18.7	92.5	7.5
P20	n.d.	95.0	28.5	57.3	14.2	96.0	4.0
P21	n.d.	26.6	29.1	64.1	6.8	97.7	2.3
P21	n.d.	42.9	30.6	60.0	9.4	96.1	3.9
P22	n.d.	38.6	20.4	57.8	21.8	87.1	13.0
P23	n.d.	9.5	21.0	73.6	5.4	99.0	1.0
P24	n.d.	44.0	20.1	70.9	8.9	96.1	3.9
P25	n.d.	41.7	24.4	66.1	9.5	97.5	2.5
P26	n.d.	32.0	30.7	65.5	3.8	99.6	0.4
P26	n.d.	3.0	32.3	66.1	1.7	100.0	0.0
P27	n.d.	22.4	29.2	63.6	7.2	97.4	2.6
P27	n.d.	40.2	37.6	56.2	6.2	97.8	2.2
P27	n.d.	34.2	27.2	67.5	5.3	98.4	1.6
P27	n.d.	22.0	52.0	46.6	1.4	99.9	0.1
P27	n.d.	64.3	28.6	68.4	3.0	98.9	1.1
P28	n.d.	42.2	24.3	72.3	3.4	99.7	0.3
P28	n.d.	48.4	38.4	58.7	2.9	99.8	0.2
			<2 μm	<63 μm	<2 mm	<0.125 mm	> 0.125 mm

Profile	[%]					
	C	H	N	TOC	TIC	TC
P01	8.6	1.8	0.0	1.6	7.0	8.6
P01	8.9	1.3	0.0	1.3	7.4	8.7
P01	8.7	1.4	0.0	0.7	7.7	8.5
P02	9.0	1.4	0.0	1.9	7.1	9.0
P03	6.4	2.5	0.2	4.0	2.4	6.4
P04	1.5	2.3	0.0	0.9	0.5	1.4
P04	1.5	2.1	0.0	0.7	0.6	1.3
P05	7.0	1.8	0.0	1.6	5.4	7.0
P06	8.8	1.1	0.0	0.5	8.6	9.2
P07	8.6	1.5	0.0	2.3	6.4	8.6
P07	7.3	1.3	0.0	0.7	6.6	7.3
P08	2.1	2.2	0.0	1.0	1.0	2.0
P09	1.3	2.1	0.0	1.0	0.2	1.2
P09	1.3	1.7	0.0	1.0	0.2	1.1
P09	1.1	1.7	0.0	0.7	0.3	0.9
P10	1.9	2.1	0.1	1.7	0.1	1.8
P10	1.9	2.1	0.0	1.6	0.1	1.7
P10	0.9	2.1	0.0	0.7	0.1	0.8
P11	3.3	2.1	0.0	0.6	2.6	3.2
P12	1.7	2.5	0.0	1.6	0.1	1.7
P13	2.0	1.9	0.0	1.7	0.1	1.8
P14	4.2	2.4	0.0	1.2	2.8	4.0
P14	6.1	1.0	0.0	0.7	5.2	5.9
P14	6.0	1.0	0.0	0.6	5.2	5.8
P15	11.3	1.4	0.2	3.4	8.0	11.5
P15	8.0	1.0	0.0	0.5	7.6	8.1
P15	8.0	1.0	0.0	0.5	7.6	8.1
P16	4.9	2.4	0.2	3.6	1.2	4.7
P17	7.7	1.0	0.1	2.0	5.7	7.6
P17	7.7	0.9	0.0	0.7	6.8	7.5
P18	8.7	2.3	0.1	2.6	6.1	8.7
P19	9.7	1.3	0.0	1.7	8.3	10.0
P20	7.6	1.6	0.2	3.3	4.3	7.5
P20	4.9	1.7	0.0	0.9	4.1	5.0
P21	4.9	1.7	0.0	0.9	4.1	5.0
P21	7.3	0.8	0.0	0.4	7.2	7.6
P22	8.7	0.7	0.0	1.0	8.1	9.1
P23	3.3	1.9	0.2	3.1	0.3	3.4
P24	7.3	1.9	0.4	5.4	2.0	7.5
P25	9.5	0.9	0.1	2.8	6.9	9.7
P26	3.3	1.8	0.2	3.0	0.5	3.6
P26	2.1	1.9	0.1	1.9	0.3	2.2
P27	6.9	1.0	0.0	1.5	5.6	7.1
P27	7.0	0.8	0.0	0.8	6.6	7.4
P27	6.3	0.9	0.0	0.7	6.0	6.7
P27	2.3	1.6	0.0	0.7	1.7	2.4
P27	8.7	0.5	0.0	0.1	8.8	9.0
P28	3.4	1.6	0.1	1.8	1.5	3.3
P28	2.1	1.6	0.0	1.0	1.0	2.0

Profile	[mg kg^{-1}]								
	Al	Ba	Ca	Cr	Cu	Fe	K	Na	Mg
P01	<0.15	84.0	3710	<0.075	<0.125	<0.075	23.5	19.7	132
P01	<0.15	84.8	3530	<0.075	0.1	0.1	25.3	28.2	124
P01	<0.15	64.1	2930	<0.075	0.2	<0.075	19.6	37.4	115
P02	<0.15	97.9	4200	<0.075	0.2	0.1	28.8	25.5	144
P03	<0.15	373.0	7570	<0.075	<0.125	0.1	104.0	50.8	382
P04	<0.15	141.0	9590	<0.075	<0.125	<0.075	76.6	39.9	442
P04	<0.15	147.0	10930	<0.075	<0.125	<0.075	83.0	57.8	451
P05	<0.15	163.0	5420	<0.075	0.1	<0.075	198.0	18.6	189
P06	<0.15	39.5	3090	<0.075	<0.125	<0.075	37.7	34.0	248
P07	<0.15	115.0	5910	<0.075	<0.125	<0.075	115.0	52.3	296
P07	<0.15	120.0	5690	<0.075	<0.125	<0.075	55.3	71.0	356
P08	<0.15	97.8	10060	<0.075	<0.125	<0.075	236.0	34.9	463
P09	<0.15	101.0	9240	<0.075	<0.125	<0.075	272.0	43.6	426
P09	0.3	109.0	9870	<0.075	<0.125	0.3	239.0	39.2	460
P09	0.3	105.0	8540	<0.075	<0.125	0.3	198.0	80.7	533
P10	<0.15	73.4	8960	<0.075	<0.125	<0.075	88.6	82.3	722
P10	<0.15	61.1	9530	<0.075	<0.125	0.1	166.0	32.7	320
P10	<0.15	107.0	10150	<0.075	<0.125	<0.075	124.0	28.9	478
P11	<0.15	111.0	8650	<0.075	0.1	<0.075	157.0	64.0	591
P12	<0.15	56.2	7270	<0.075	<0.125	0.1	111.0	126.0	641
P13	<0.15	69.5	7870	<0.075	0.1	0.1	76.0	72.6	705
P14	<0.15	118.0	5760	<0.075	0.1	<0.075	283.0	52.3	236
P14	<0.15	100.0	4230	<0.075	0.1	1.7	93.2	60.4	214
P14	<0.15	77.9	3790	<0.075	0.1	0.1	133.0	37.7	207
P15	0.3	71.7	3500	<0.075	0.2	0.2	76.6	44.9	137
P15	0.2	84.6	3400	<0.075	0.1	1.4	32.2	44.5	136
P15	0.2	89.6	3490	<0.075	<0.125	0.2	52.8	165.0	294
P16	<0.15	76.6	9280	<0.075	<0.125	0.1	101.0	65.2	366
P17	<0.15	79.4	4240	<0.075	0.1	0.2	85.6	22.3	140
P17	0.2	80.0	2910	<0.075	<0.125	0.1	29.0	13.3	65
P18	<0.15	102.0	3810	<0.075	<0.125	0.1	192.0	65.0	157
P19	0.6	22.6	2510	<0.075	0.2	0.3	35.1	118.0	111
P20	0.2	40.9	6000	<0.075	0.1	0.1	187.0	82.3	263
P20	<0.15	39.4	7130	<0.075	<0.125	<0.075	108.0	54.6	258
P21	<0.15	194.0	4550	<0.075	<0.125	0.1	99.7	41.6	347
P21	<0.15	80.3	3240	<0.075	<0.125	0.1	109.0	55.9	340
P22	<0.15	81.3	2900	<0.075	<0.125	<0.075	114.0	21.2	191
P23	<0.15	94.4	9810	<0.075	<0.125	0.1	98.6	41.9	292
P24	0.4	128.0	8050	<0.075	<0.125	0.3	105.0	108.0	381
P25	0.2	55.8	3710	<0.075	0.2	0.3	364.0	34.9	181
P26	0.2	84.3	8280	<0.075	<0.125	0.2	600.0	52.3	509
P26	<0.15	109.0	9280	<0.075	<0.125	<0.075	139.0	59.6	375
P27	<0.15	69.8	4280	<0.075	<0.125	0.2	558.0	45.2	174
P27	0.4	96.0	4080	<0.075	<0.125	0.3	355.0	28.8	128
P27	0.8	123.0	4360	<0.075	<0.125	0.6	285.0	39.9	135
P27	<0.15	128.0	8710	<0.075	<0.125	0.1	1020.0	104.0	706
P27	0.4	19.3	2590	<0.075	<0.125	0.3	335.0	51.5	225
P28	<0.15	71.9	8490	<0.075	<0.125	0.1	172.0	28.3	293
P28	<0.15	66.4	7840	<0.075	0.4	0.2	220.0	14.5	271

Profile	[mg kg-1]							
	Mn	Mo	Ni	P	Pb	S	Ti	Zn
P01	<0.025	<0.15	<0.1	<0.25	<0.15	2.2	<0.225	<0.125
P01	<0.025	<0.15	<0.1	0.5	<0.15	4.1	<0.225	<0.125
P01	<0.025	<0.15	<0.1	0.3	<0.15	10.6	<0.225	<0.125
P02	<0.025	<0.15	<0.1	0.4	<0.15	4.0	<0.225	<0.125
P03	0.1	<0.15	<0.1	<0.25	<0.15	7.3	<0.225	<0.125
P04	<0.025	<0.15	<0.1	<0.25	<0.15	3.0	<0.225	<0.125
P04	0.0	<0.15	<0.1	<0.25	<0.15	1.0	<0.225	<0.125
P05	0.0	<0.15	<0.1	0.3	<0.15	4.4	<0.225	<0.125
P06	0.0	<0.15	<0.1	0.3	<0.15	8.4	<0.225	<0.125
P07	0.2	<0.15	<0.1	0.3	<0.15	6.9	<0.225	<0.125
P07	<0.025	<0.15	<0.1	<0.25	<0.15	9.6	<0.225	<0.125
P08	<0.025	<0.15	<0.1	<0.25	<0.15	2.1	<0.225	<0.125
P09	0.0	<0.15	<0.1	0.4	<0.15	3.0	<0.225	<0.125
P09	0.3	<0.15	<0.1	0.5	<0.15	1.3	<0.225	<0.125
P09	0.0	<0.15	<0.1	0.4	<0.15	4.0	<0.225	<0.125
P10	0.1	<0.15	<0.1	0.4	<0.15	8.3	<0.225	<0.125
P10	0.0	<0.15	<0.1	0.4	<0.15	6.4	<0.225	<0.125
P10	0.1	<0.15	<0.1	0.3	<0.15	1.7	<0.225	<0.125
P11	0.0	0.6	<0.1	<0.25	<0.15	4.2	<0.225	<0.125
P12	0.0	<0.15	<0.1	0.4	<0.15	7.4	<0.225	<0.125
P13	0.2	<0.15	<0.1	0.3	<0.15	6.9	<0.225	<0.125
P14	0.0	<0.15	<0.1	<0.25	<0.15	3.9	<0.225	<0.125
P14	0.2	<0.15	<0.1	0.7	<0.15	14.7	<0.225	<0.125
P14	0.2	<0.15	<0.1	0.3	<0.15	22.7	<0.225	<0.125
P15	0.1	<0.15	<0.1	0.6	<0.15	7.4	<0.225	<0.125
P15	0.1	<0.15	<0.1	0.7	0.5	14.0	<0.225	<0.125
P15	0.1	<0.15	<0.1	0.4	<0.15	17.8	<0.225	<0.125
P16	0.1	<0.15	<0.1	0.4	<0.15	8.8	<0.225	<0.125
P17	0.1	<0.15	<0.1	0.6	<0.15	6.5	<0.225	<0.125
P17	<0.025	<0.15	<0.1	0.4	<0.15	2.4	<0.225	<0.125
P18	<0.025	<0.15	<0.1	<0.25	<0.15	7.1	<0.225	<0.125
P19	0.0	<0.15	<0.1	0.6	<0.15	34.1	<0.225	<0.125
P20	0.3	<0.15	<0.1	0.5	<0.15	9.9	<0.225	<0.125
P20	0.5	<0.15	<0.1	0.4	<0.15	5.3	<0.225	<0.125
P21	0.1	<0.15	<0.1	0.5	<0.15	2.5	<0.225	<0.125
P21	0.0	<0.15	<0.1	0.3	<0.15	4.0	<0.225	<0.125
P22	0.3	<0.15	<0.1	0.7	<0.15	5.4	<0.225	<0.125
P23	0.9	<0.15	<0.1	0.8	<0.15	4.8	<0.225	<0.125
P24	0.4	<0.15	<0.1	1.1	<0.15	11.9	<0.225	<0.125
P25	2.4	<0.15	<0.1	2.6	<0.15	12.2	<0.225	<0.125
P26	2.0	<0.15	<0.1	2.3	<0.15	8.3	<0.225	<0.125
P26	0.0	<0.15	<0.1	0.3	<0.15	<0.2	<0.225	<0.125
P27	0.3	<0.15	<0.1	1.3	<0.15	3.5	<0.225	<0.125
P27	0.1	<0.15	<0.1	0.9	<0.15	1.7	<0.225	<0.125
P27	0.1	<0.15	<0.1	0.9	<0.15	6.4	<0.225	<0.125
P27	<0.025	<0.15	<0.1	0.3	1.6	4.4	<0.225	<0.125
P27	0.1	<0.15	<0.1	0.3	<0.15	15.2	<0.225	<0.125
P28	1.4	<0.15	<0.1	1.0	<0.15	3.8	<0.225	<0.125
P28	0.0	<0.15	<0.1	0.5	0.2	1.0	<0.225	0.6

MIX
Papier aus verantwortungsvollen Quellen
Paper from responsible sources
FSC® C105338

If you have any concerns about our products,
you can contact us on
ProductSafety@springernature.com

In case Publisher is established outside the EU,
the EU authorized representative is:
Springer Nature Customer Service Center GmbH
Europaplatz 3, 69115 Heidelberg, Germany

Printed by Libri Plureos GmbH
in Hamburg, Germany